# Make Music Count

## Graphing Edition

Printed in the United States of America
First Printing, 2015
Make Music Count, LLC Atlanta, GA
www.MakeMusicCount.com

# Contents

## Abstract

My life thus far has been filled with moments and events that prepared me for both math and music separately. But all of those moments were needed in order for me to combine math and music at one specific time. The two came together for me when I was asked to coordinate the music for my good friend Orson's wedding. While I was playing the piano at his rehearsal Adam came up and asked me "Can you explain that?" I looked down at the piano and said "Yeah I think so." I knew that since he played the bass he didn't know anything about piano music theory so I explained my chords the only way I knew how and that was through using math. This was when I realized that I made sense out of the musical notes and chords that I played from only using a math correlation. But it didn't stop there because I began to wonder if there were others that were aspiring pianist that didn't know any music theory or how to read music. Could they also learn how to play simply from applying mathematics? My interests in math and music began back home in Connecticut growing up. I attended The Artists Collective in Hartford CT a performing arts center, where I studied classical and jazz piano and also played the trumpet in the jazz orchestra. I had the privilege of being able to learn from Jackie McLean and other phenomenal jazz musicians. I also began to develop the ability to play music by ear on the piano from attending church regularly and wanting to learn how to play what I heard in service. So my musical abilities were developing rapidly during middle and high school.

But I wouldn't have been able to explain any of this to anyone else until I developed mathematically. Math had always been strength of mine, which is why I attended the Greater Hartford Academy of Mathematics and Science in Hartford CT. But even though I spent all of my time with math in high school it wasn't until I attended Morehouse College and read "The Eight" By Katherine Neville that I fully embraced Mathematics as a major. In the book there was a scene where a pianist took the movements made by each chess player during their match and created a musical piece from it. I found this extremely interesting that something as quantitative as chess movements could somehow turn into music. This curiosity sparked my decision to take on Mathematics as a major with the hope of one day making a connection to music. What I realized was that it was the math application that resulted in music that caused me to accept the idea of pursuing mathematics. Being able to produce music from only using math made doing math seem friendlier. For me mathematics in previous setting was taught as a concept that was hard to grasp and understand. Before even entering a class most students have already made up in their mind that they cannot and will not figure out math creating a defeated mentality that is impossible for teachers to break down. So once I realized that I could teach others music with math I decided to think about it in a more educational way: What if I could teach math with music? I then began to derive a method in which math lessons were taught through music and The Musical Number Line was created.

In this journal you will walk through piano lessons of great hip hop songs that will be derived from mathematical application. This journal is your guide to applying mathematics to learn how to play the piano while also strengthening your understanding of basic algebra concepts. Every song in this journal will be learned by applying only mathematics.

Here is a picture of how the piano is normally taught where we are asked to view the notes as if we were looking at an actual piano.

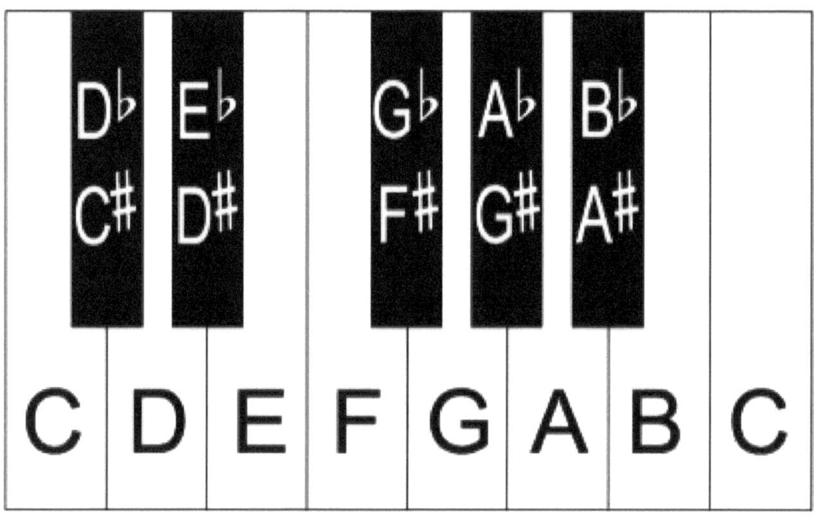


We will now look at the piano in a completely different manner; as a number line. Each musical note will now represent a point on a line. And the movement from one point on the line to another will signify the movement of playing one musical note to another. Here the concern is only the numerical distance between musical notes. By viewing the piano notes in this manner we will remove the music theory out of the music and be able to directly apply the notes on the number line to playing on the piano.

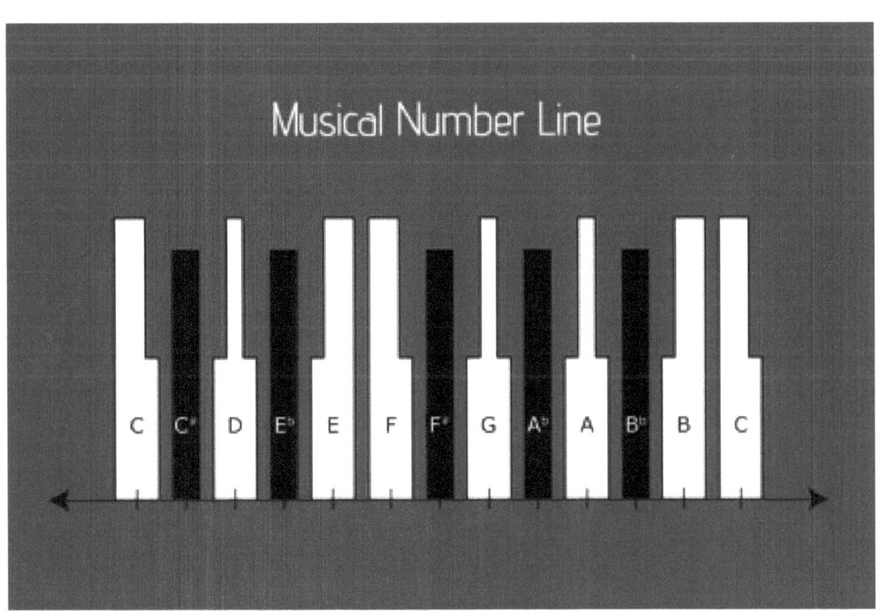
Musical Number Line
C
C#
D
Eb
E
F
F#
G
Ab
A
Bb
B
C

# 1 Half Steps

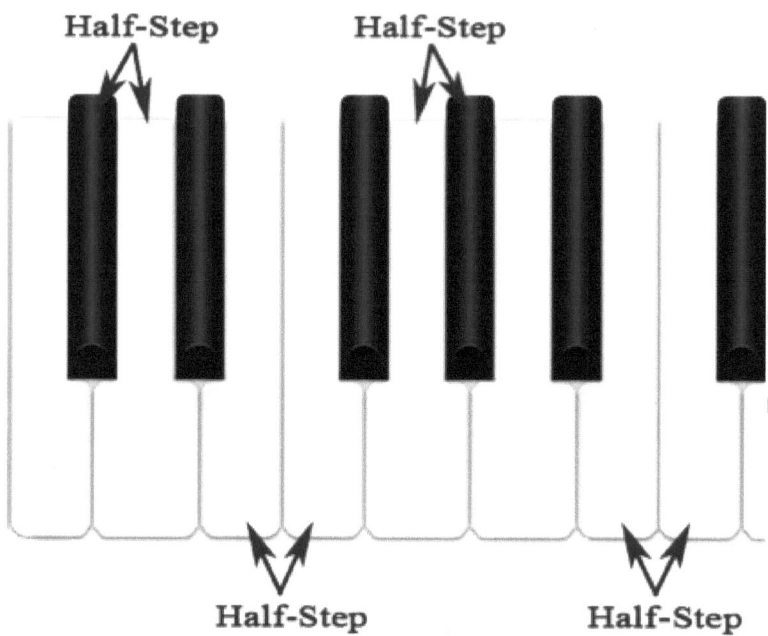


A half step in music is when you play one note on the piano and then move up or down to the very next note to play. For example if you played the note C, a half step up would be $C\sharp$, while a half step down from C would be the note B. On our musical number line we will represent a half step movement as 1/2. From this conclusion we see that the distance between each note on the Musical Number line is $\frac{1}{2}$.

Examples:

1. G + $\frac{1}{2}$ = $A\flat$
2. $B\flat$ − $\frac{1}{2}$ = A

# Half Step Practice

C $+ \frac{1}{2} =$ ____________

A $+ \frac{1}{2} =$ ____________

C♯ $+ \frac{1}{2} =$ ____________

E♭ $+ \frac{1}{2} =$ ____________

D♭ $+ \frac{1}{2} =$ ____________

G $+ \frac{1}{2} =$ ____________

E $+ \frac{1}{2} =$ ____________

F♯ $+ \frac{1}{2} =$ ____________

B♭ $+ \frac{1}{2} =$ ____________

B $+ \frac{1}{2} =$ ____________

G♭ $+ \frac{1}{2} =$ ____________

F $+ \frac{1}{2} =$ ____________

A♯ $+ \frac{1}{2} =$ ____________

D♯ $+ \frac{1}{2} =$ ____________

F $- \frac{1}{2} =$ ____________

B $- \frac{1}{2} =$ ____________

A♭ $- \frac{1}{2} =$ ____________

C $- \frac{1}{2} =$ ____________

C♯ $- \frac{1}{2} =$ ____________

E♭ $- \frac{1}{2} =$ ____________

G♭ $- \frac{1}{2} =$ ____________

A $- \frac{1}{2} =$ ____________

B♭ $- \frac{1}{2} =$ ____________

E $- \frac{1}{2} =$ ____________

D♯ $- \frac{1}{2} =$ ____________

A♯ $- \frac{1}{2} =$ ____________

F♯ $- \frac{1}{2} =$ ____________

G $- \frac{1}{2} =$ ____________

## 2 Whole Steps

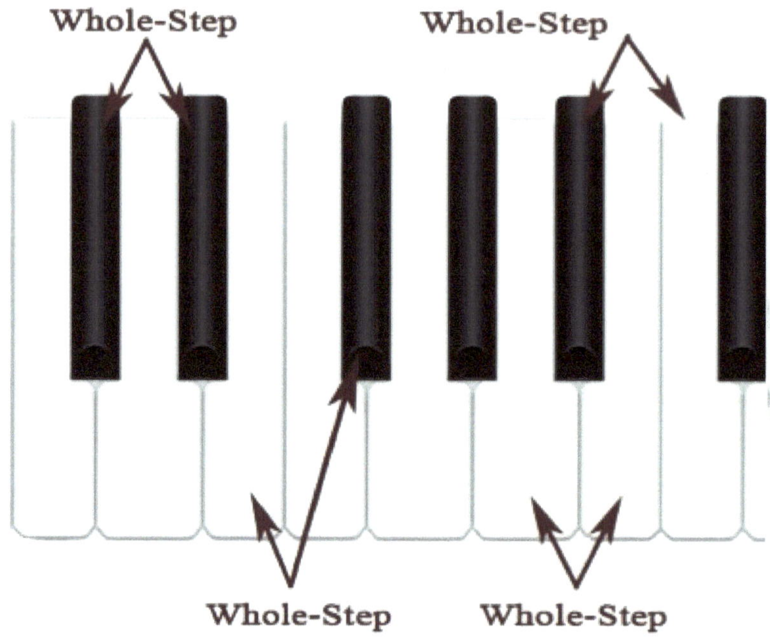


In music there are also whole steps. A whole step is when you play one note and you skip the very next note to play the following note. A whole step is also two half steps from the original starting note. An example of a whole step is if you played the note F and then moved and played the note G. Notice that in order to move from F to G the note F♯ must be skipped. We will define whole steps as 1.

Examples:

1. E + 1 = F♯

2. G + 2 = B

# Whole Step Practice

C + 1 = __________

A + 1 = __________

C♯ + 1 = __________

E♭ + 1 = __________

D♭ + 1 = __________

G + 2 = __________

E + 2 = __________

F♯ + 2 = __________

B♭ + 2 = __________

B + 5 = __________

G♭ + 1 = __________

F + 2 =

A♯ + 4 = __________

D♯ + 3 = __________

F - 1 = __________

B - 1 = __________

A♭ - 1 = __________

C - 1 = __________

C♯ - 1 = __________

E♭ - 2 = __________

G♭ - 2 = __________

A - 2 = __________

B♭ - 2 = __________

E - 5 = __________

D♯ - 1 = __________

A♯ - 2 = __________

F♯ - 4 = __________

G - 3 = __________

# 3 Major Scales

Now that we can see how math is used to show the distance between musical notes, we can use the same application for deriving musical scales. Scales have a specific distance between each of the eight notes that make up that scale. With this knowledge we can derive any type of scale using math.

Let's take the major scales as an example. A major scale has eight notes, and when played it sounds "happy." Every note on the musical number line can serve as the beginning of a major scale, which means that there are twelve major scales to learn as an aspiring pianist. But here is the good news, each major scale uses the same sequence of math distances regardless of what note it starts on.

Here is the sequence to derive any major scale:

Major Scale = $1,1,\frac{1}{2}, 1, 1, 1, \frac{1}{2}$

The numbers in the sequence represent the distance between each note of the major scale. Let's apply this sequence to derive the C Major Scale.But the way we are going to derive the C Major scale is through graphing. The Y axis represents the musical notes on the piano while the X axis represents the order in which you would play the musical notes.

Example 1: (1, F♯)

Plotting this means you would play the musical note F♯ on the piano first.

Example 2: Plotting the point (3, D) means you will play the musical note D 3rd in the order of notes to play.

Lets walk through apply this method of graphing to the C Major scale.

1. There are 8 notes to play in any Major scale.

2. We can derive the 8 notes to play by using the Major scale sequence 1,1,$\frac{1}{2}$, 1, 1, 1, $\frac{1}{2}$

3. C Major Scale notes = C, D, E, F, G, A, B, C

4. Now we can create the eight points to graph.

5. (1,C); (2,D); (3,E); (4,F); (5,G); (6,A); (7,B); (8,2C)
The X value represents the order in which the note is played and the Y value represents the note to be played. For example (2,D) means that the musical note D is played second.

6. Once the points are plotted, play each note on the piano in the order

# C Major Scale

Major Scale = {1, 1, ½, 1, 1, 1, ½}

| C | | D | | E | | F | | G | | A | | B | | C |
|---|---|---|---|---|---|---|---|---|---|---|---|---|---|---|
| | 1 | | 1 | | ½ | | 1 | | 1 | | 1 | | ½ | |

# C Major Scale

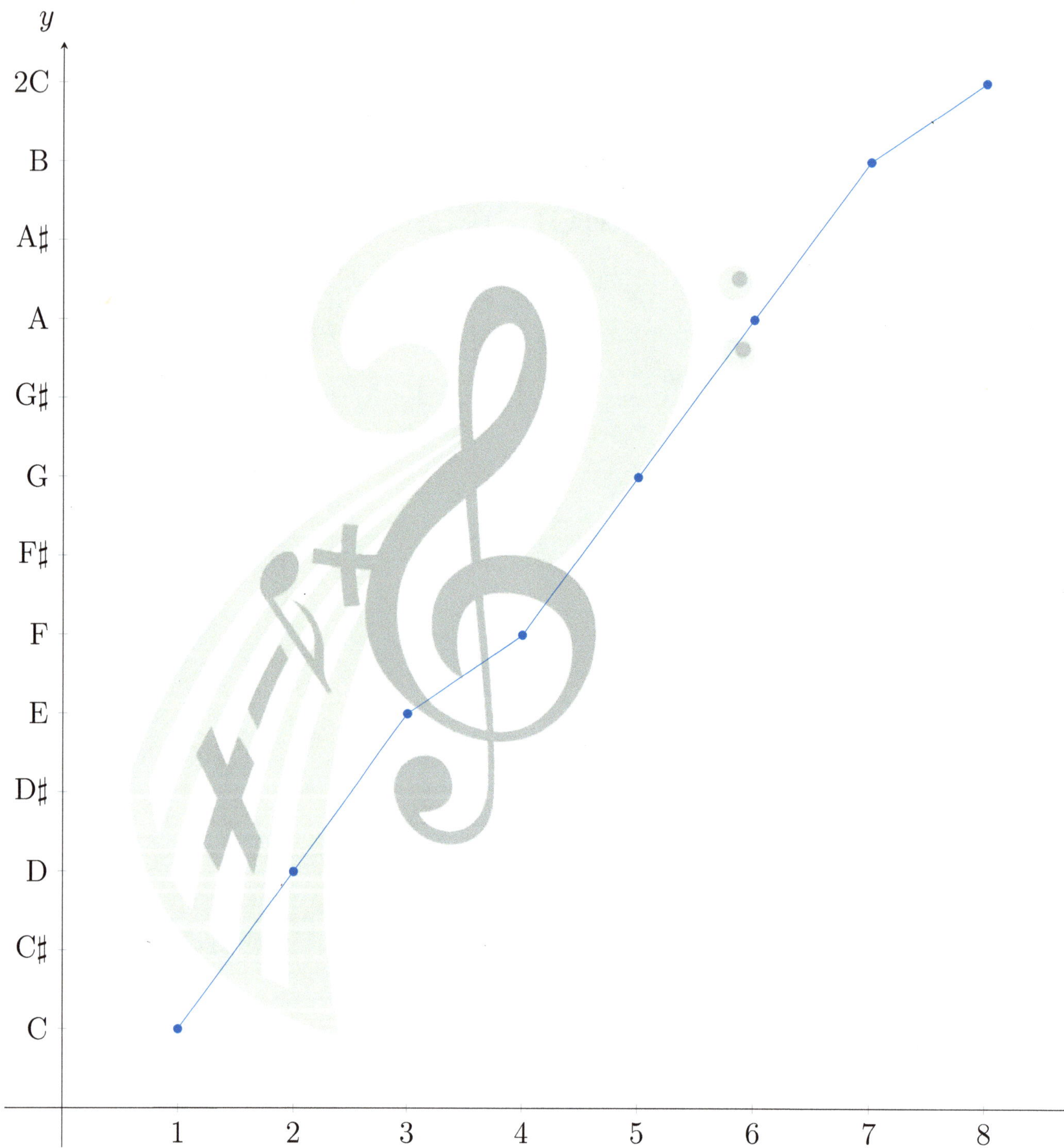

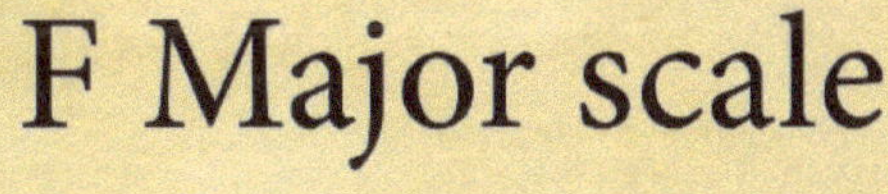


Major Scale = {1, 1, ½, 1, 1, 1, ½}

F ___ ___ B♭ ___ ___ E ___

1 1 ½ 1 1 1 ½

# F Major Scale

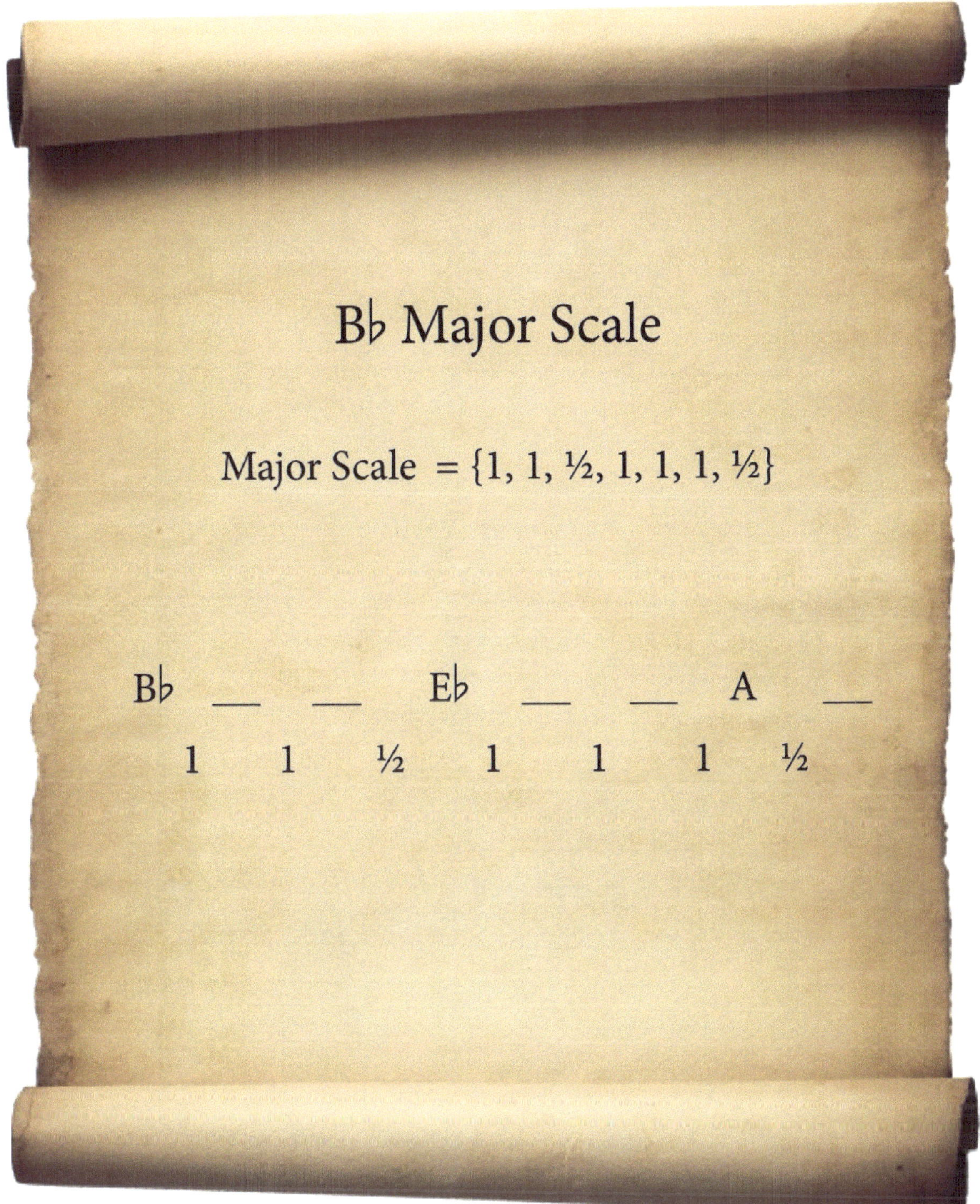

# B♭ Major Scale

Major Scale = {1, 1, ½, 1, 1, 1, ½}

| B♭ | ___ | ___ | E♭ | ___ | ___ | A | ___ |
|---|---|---|---|---|---|---|---|
| | 1 | 1 | ½ | 1 | 1 | 1 | ½ |

# B♭ Major Scale

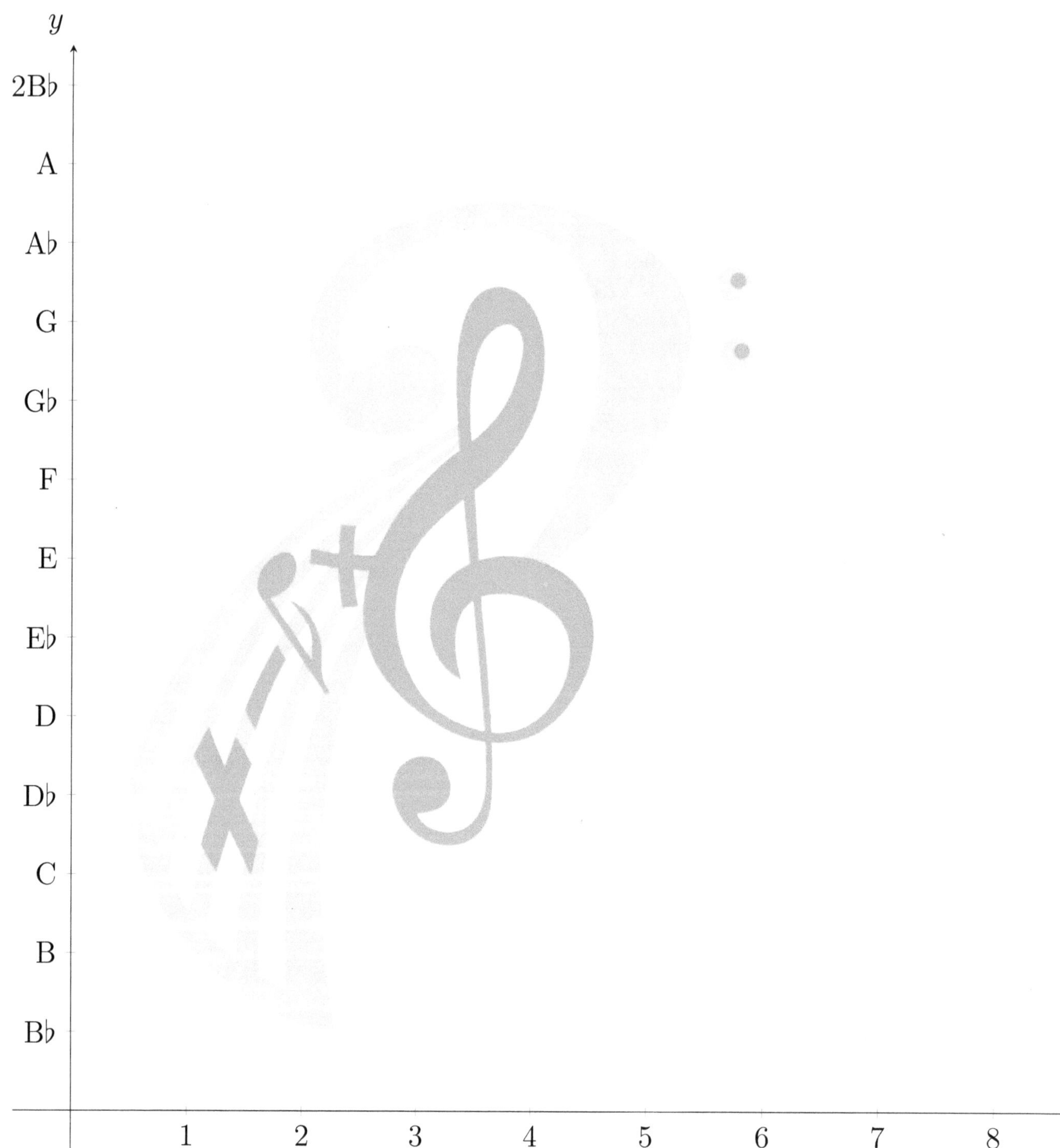

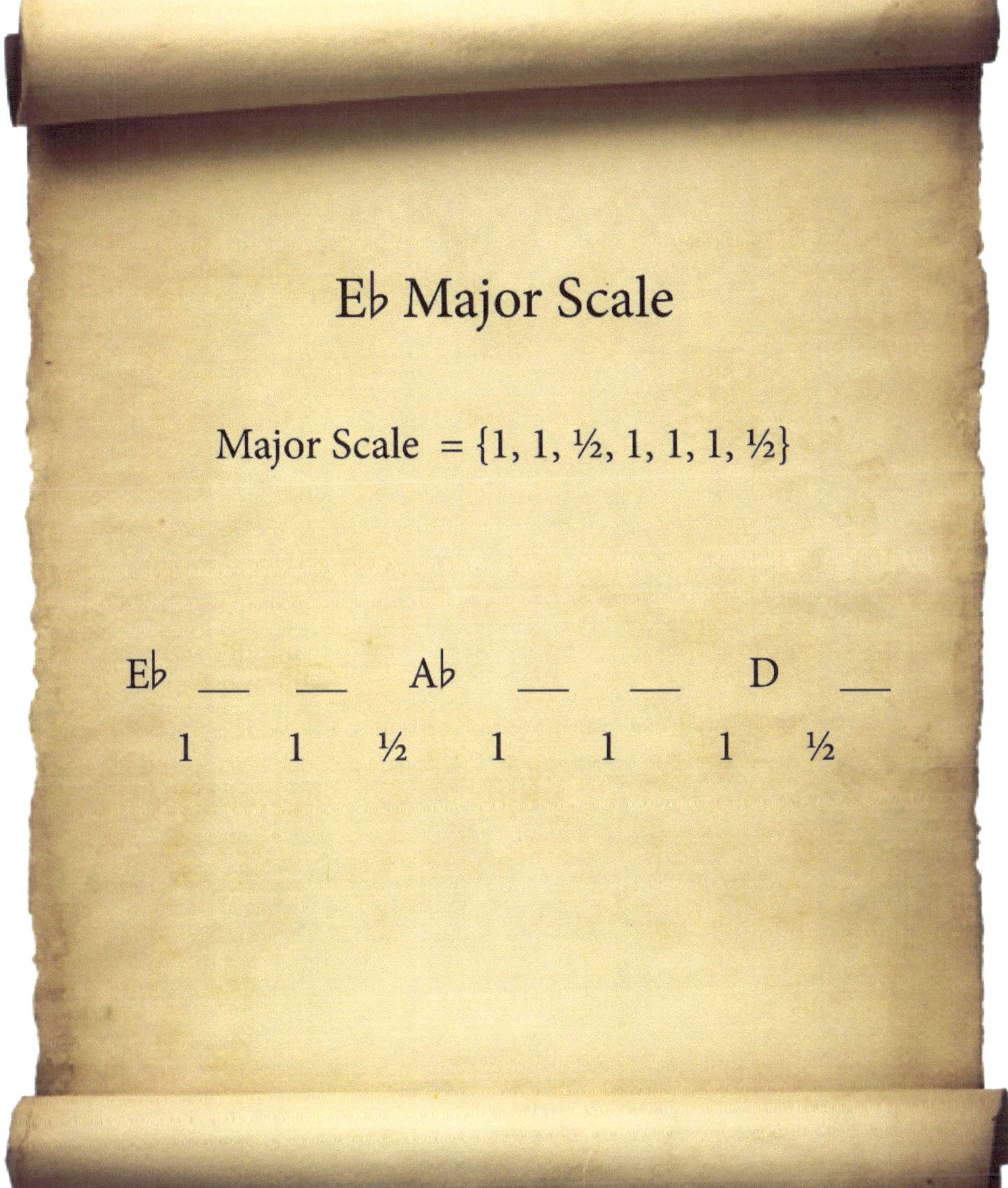

# E♭ Major Scale

Major Scale = {1, 1, ½, 1, 1, 1, ½}

E♭ ___ ___ A♭ ___ ___ D ___

1 1 ½ 1 1 1 ½

# E♭ Major Scale

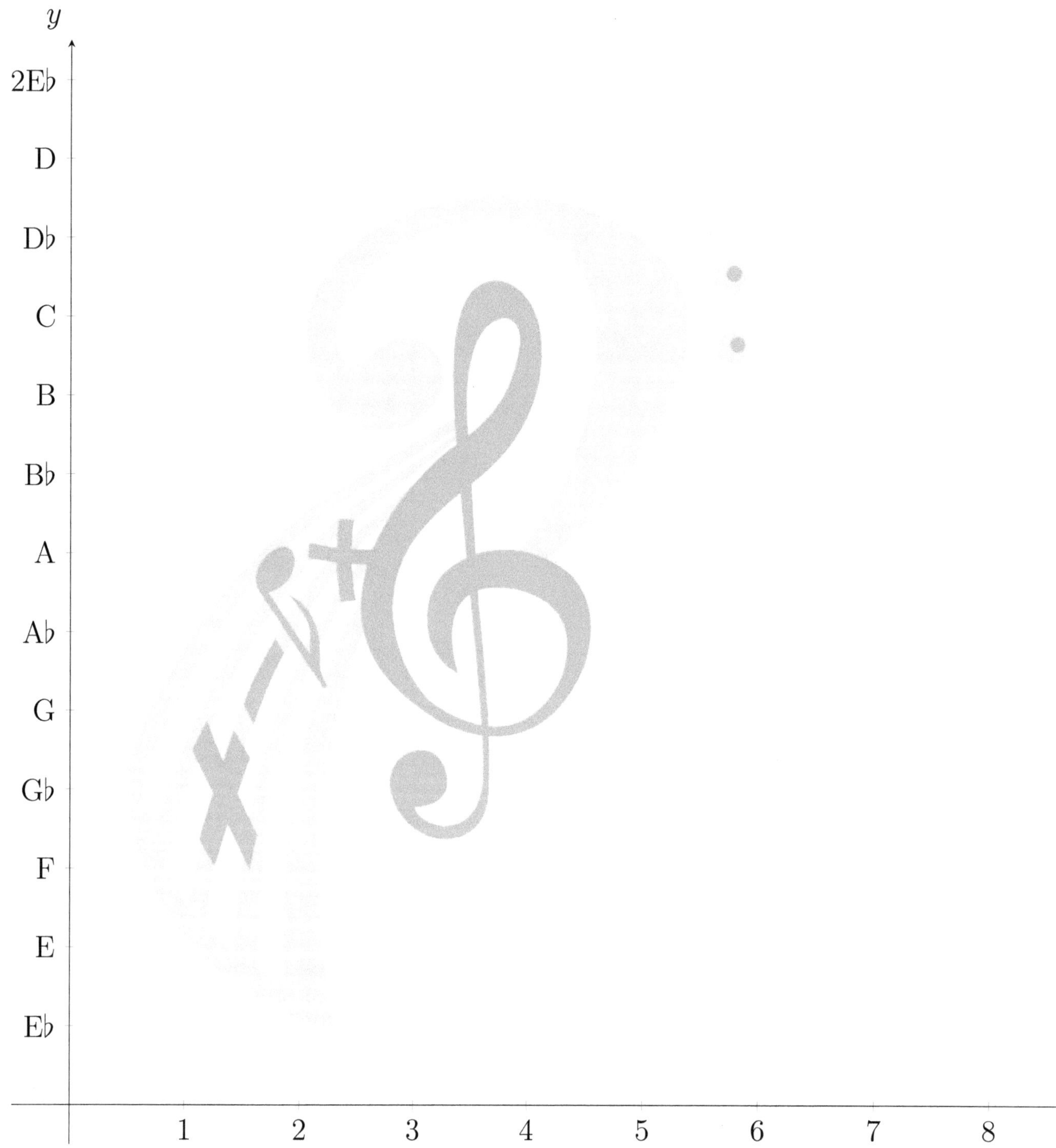

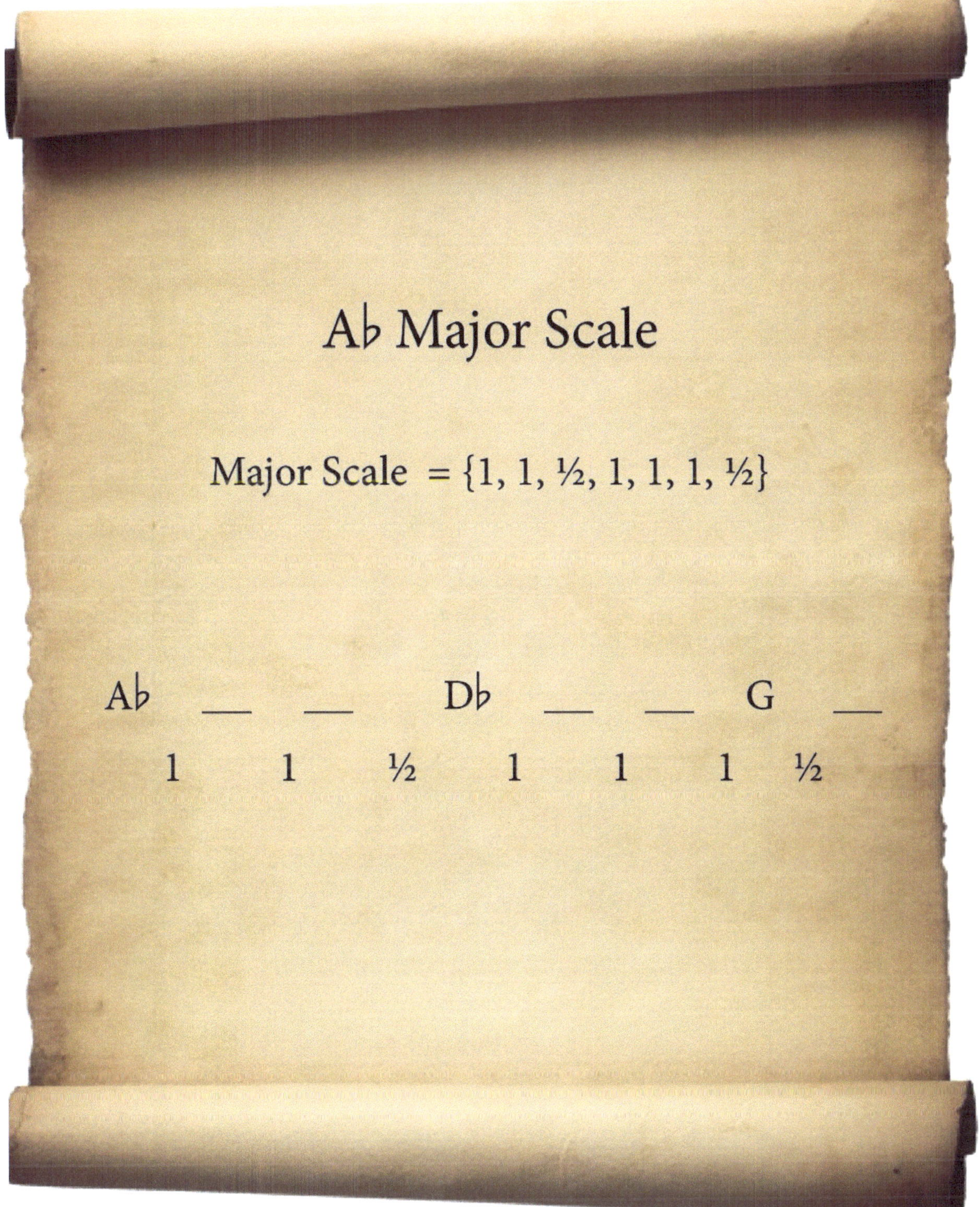

# A♭ Major Scale

Major Scale = {1, 1, ½, 1, 1, 1, ½}

| A♭ | | __ | | __ | | D♭ | | __ | | __ | | G | | __ |
|---|---|---|---|---|---|---|---|---|---|---|---|---|---|---|
| | 1 | | 1 | | ½ | | 1 | | 1 | | 1 | | ½ | |

# A♭ Major Scale

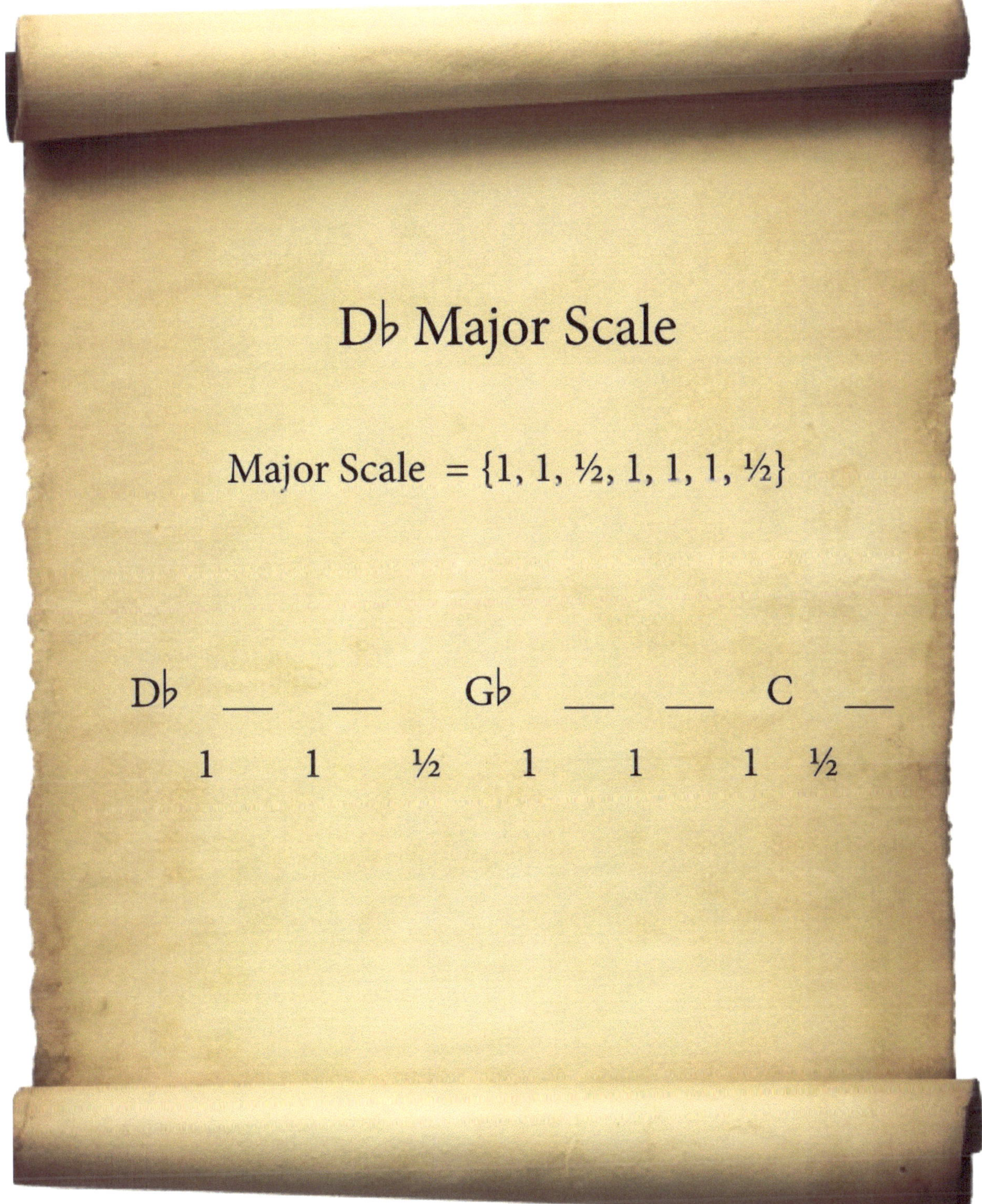
D♭ Major Scale
Major Scale = {1, 1, ½, 1, 1, 1, ½}
D♭ ___ ___ G♭ ___ ___ C ___
1 1 ½ 1 1 1 ½

# D♭ Major Scale

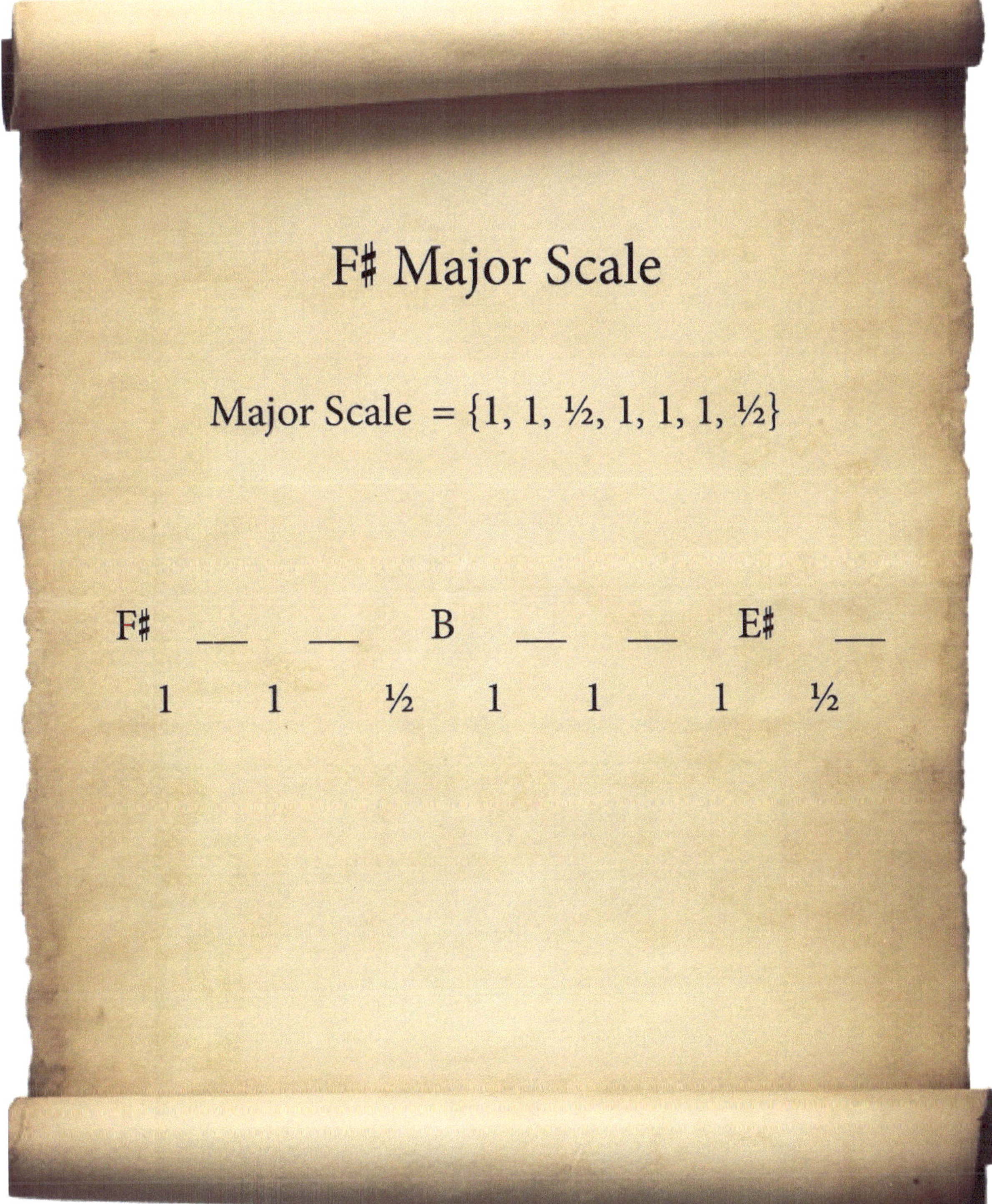
F♯ Major Scale
Major Scale = {1, 1, ½, 1, 1, 1, ½}
F♯ __ __ B __ __ E♯ __
1 1 ½ 1 1 1 ½

# F♯ Major Scale

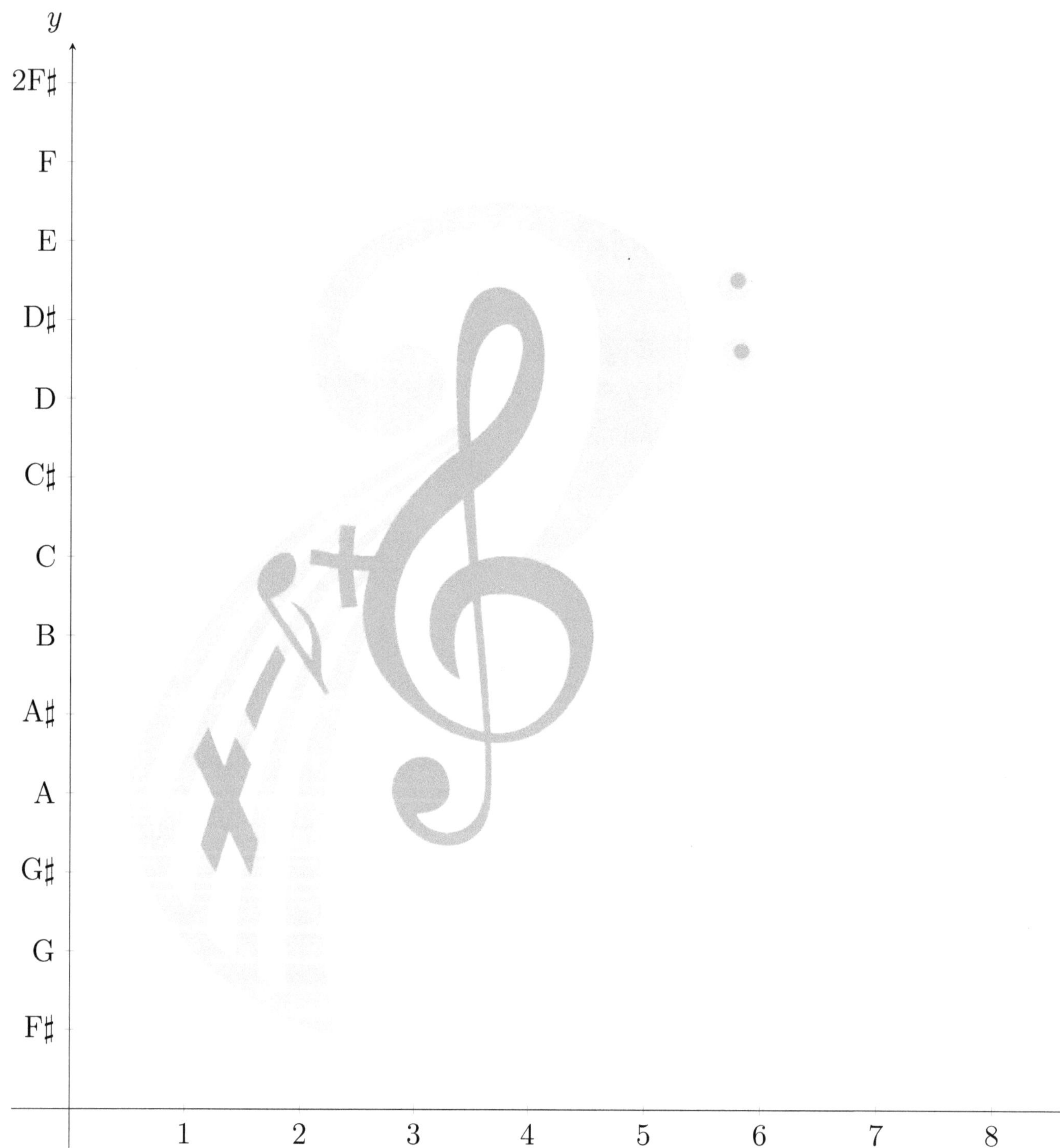

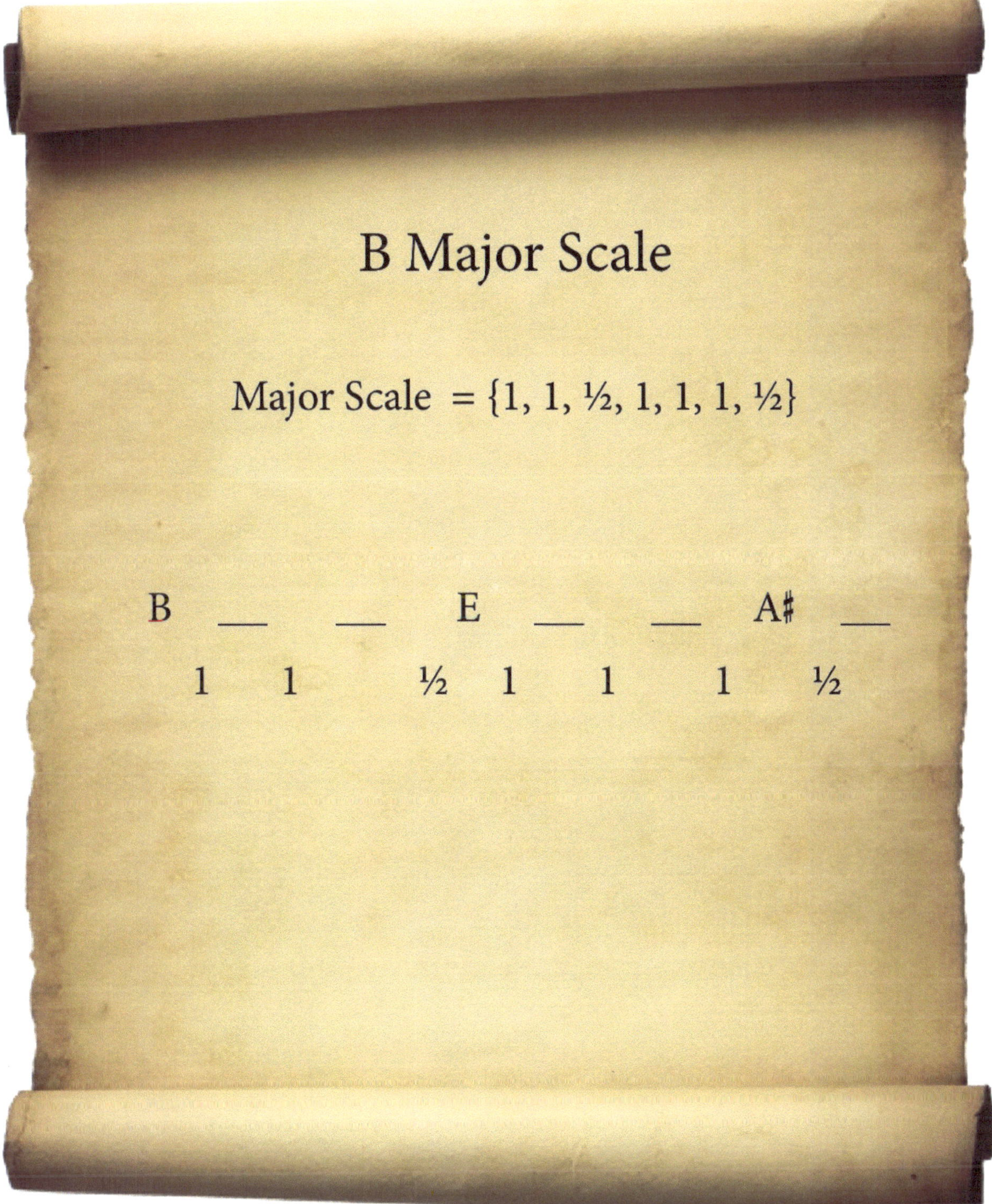

# B Major Scale

Major Scale = {1, 1, ½, 1, 1, 1, ½}

B ___ ___ E ___ ___ A♯ ___

1 1 ½ 1 1 1 ½

# B Major Scale

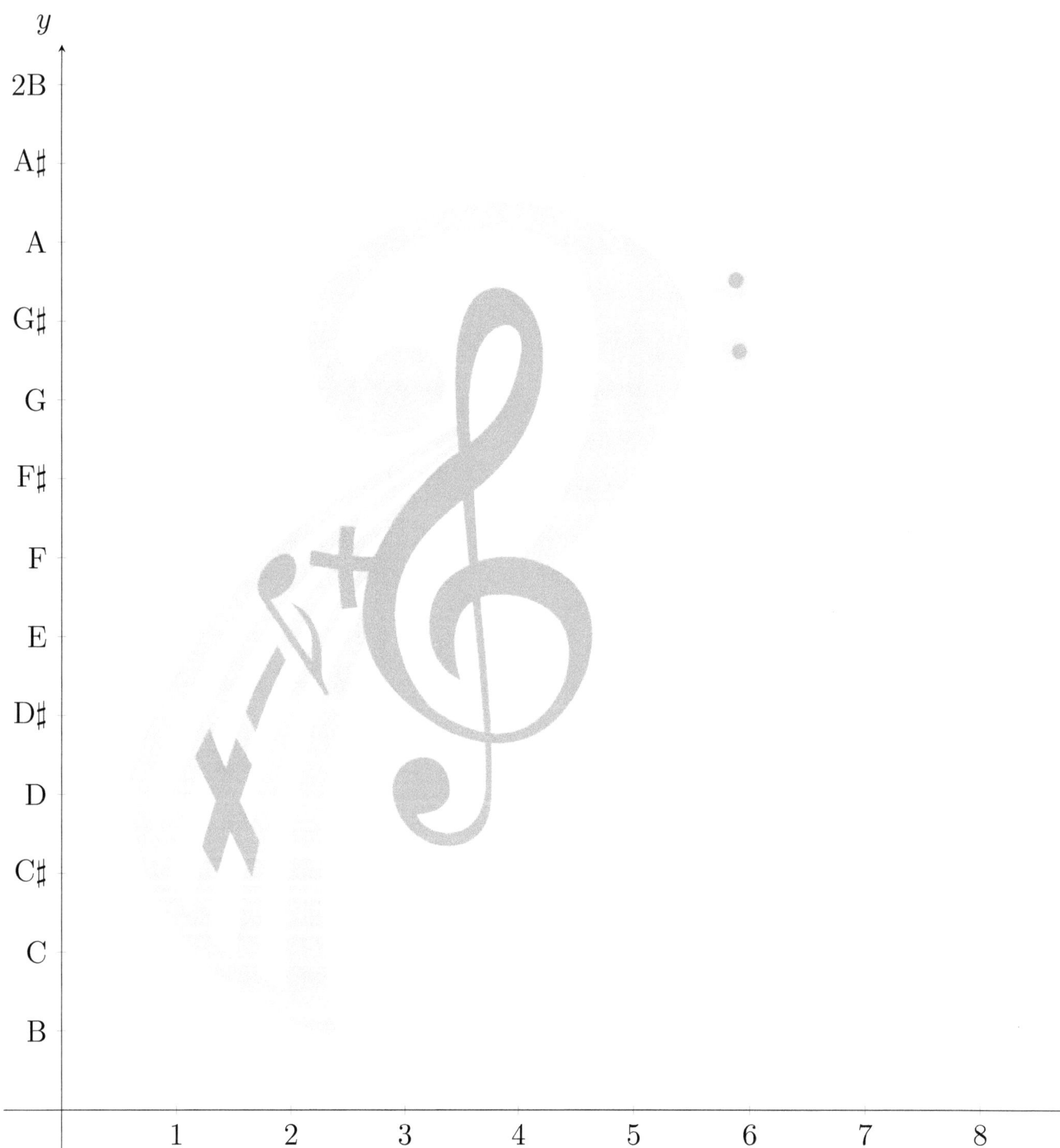

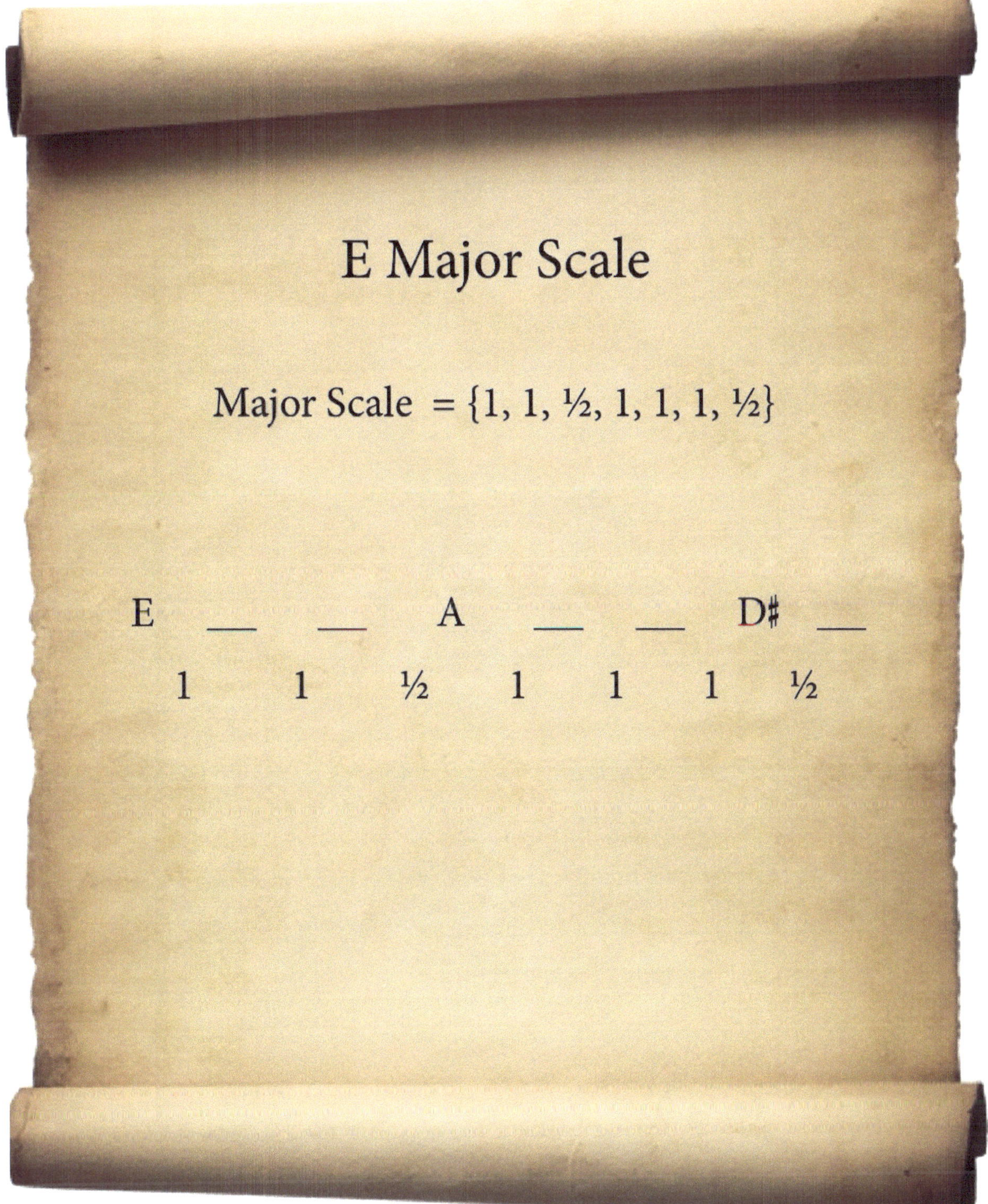
E Major Scale
Major Scale = {1, 1, ½, 1, 1, 1, ½}
E ___ ___ A ___ ___ D♯ ___
1 1 ½ 1 1 1 ½

# E Major Scale

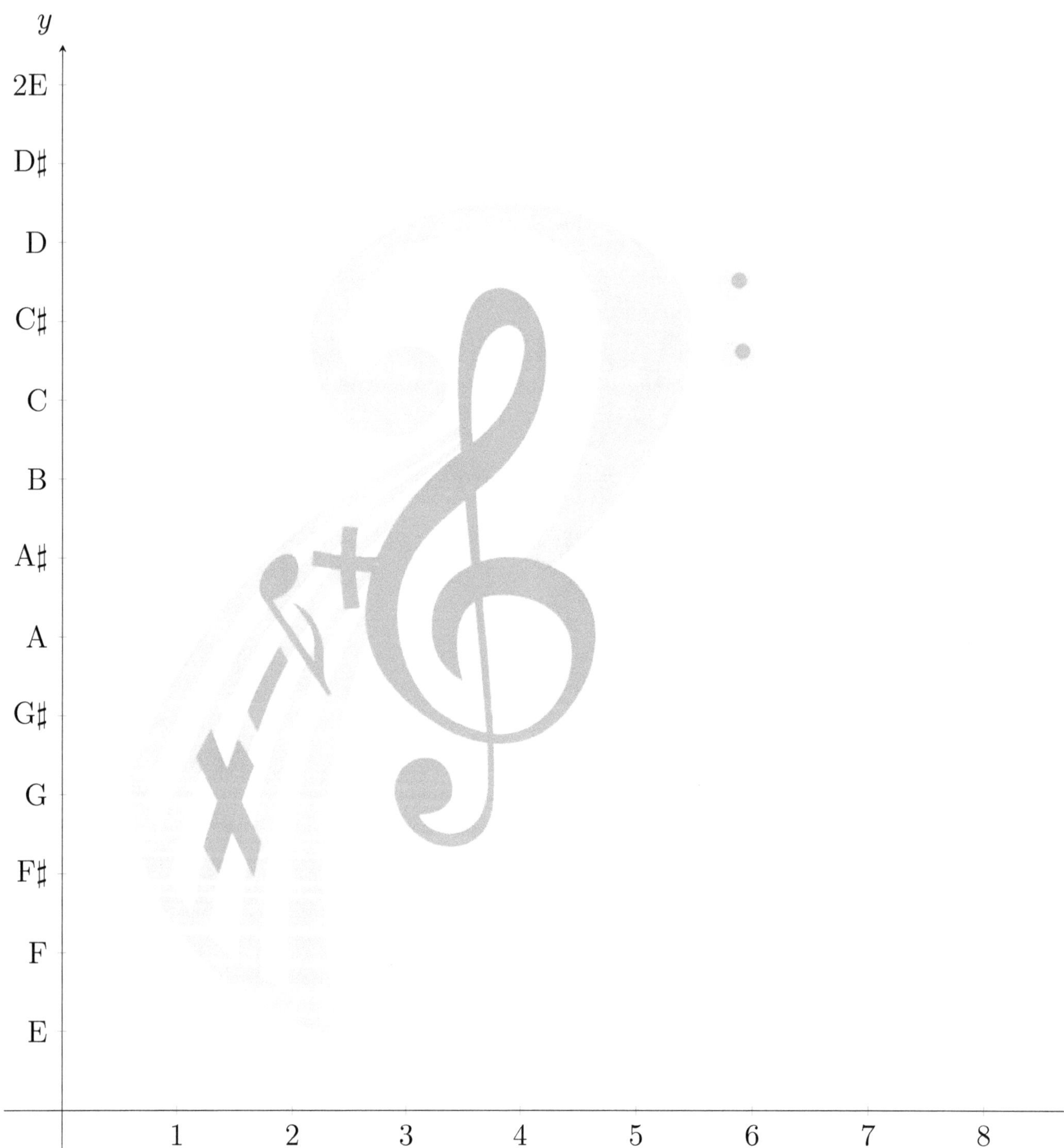

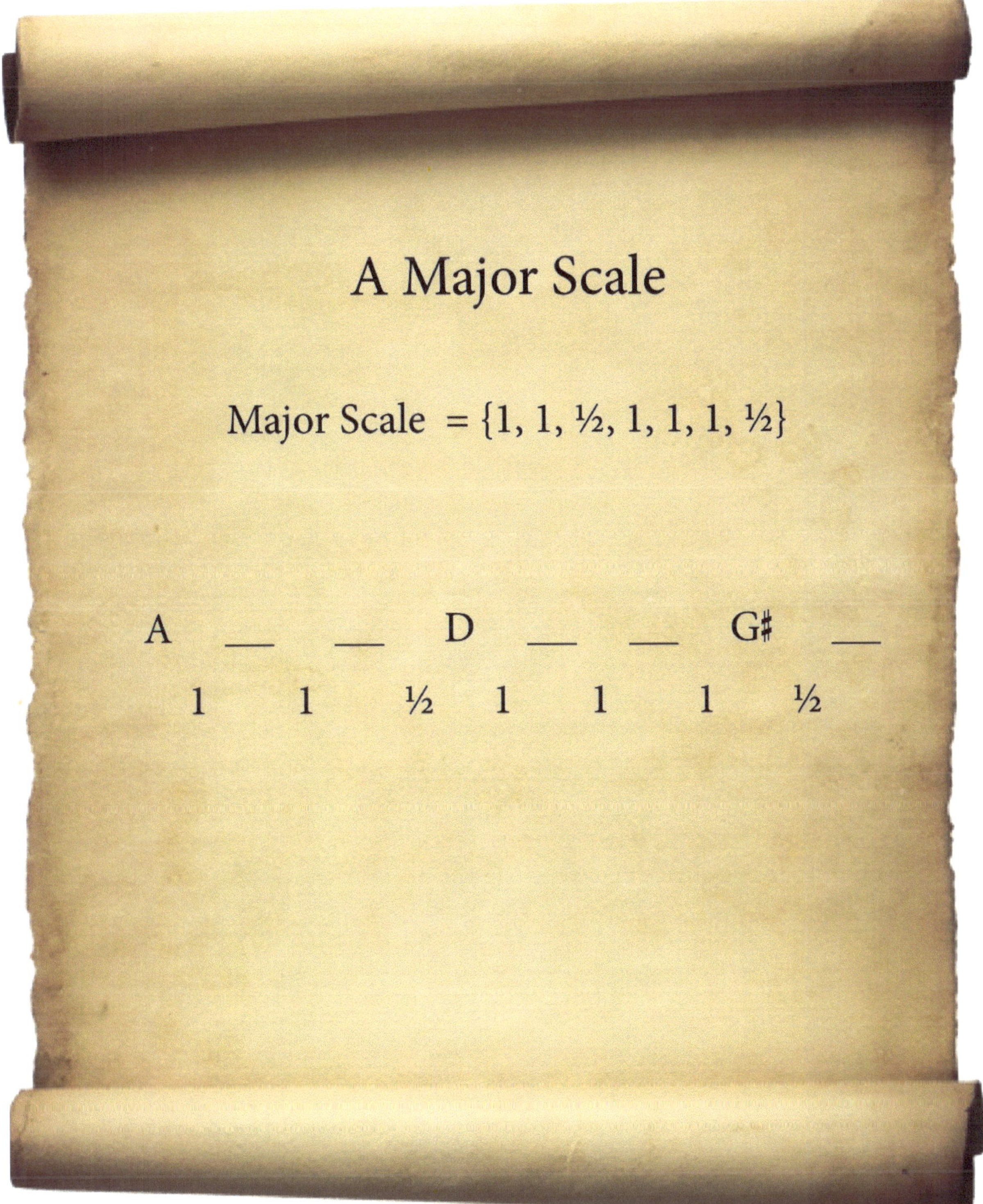
A Major Scale
Major Scale = {1, 1, ½, 1, 1, 1, ½}
A __ __ D __ __ G♯ __
1 1 ½ 1 1 1 ½

# A Major Scale

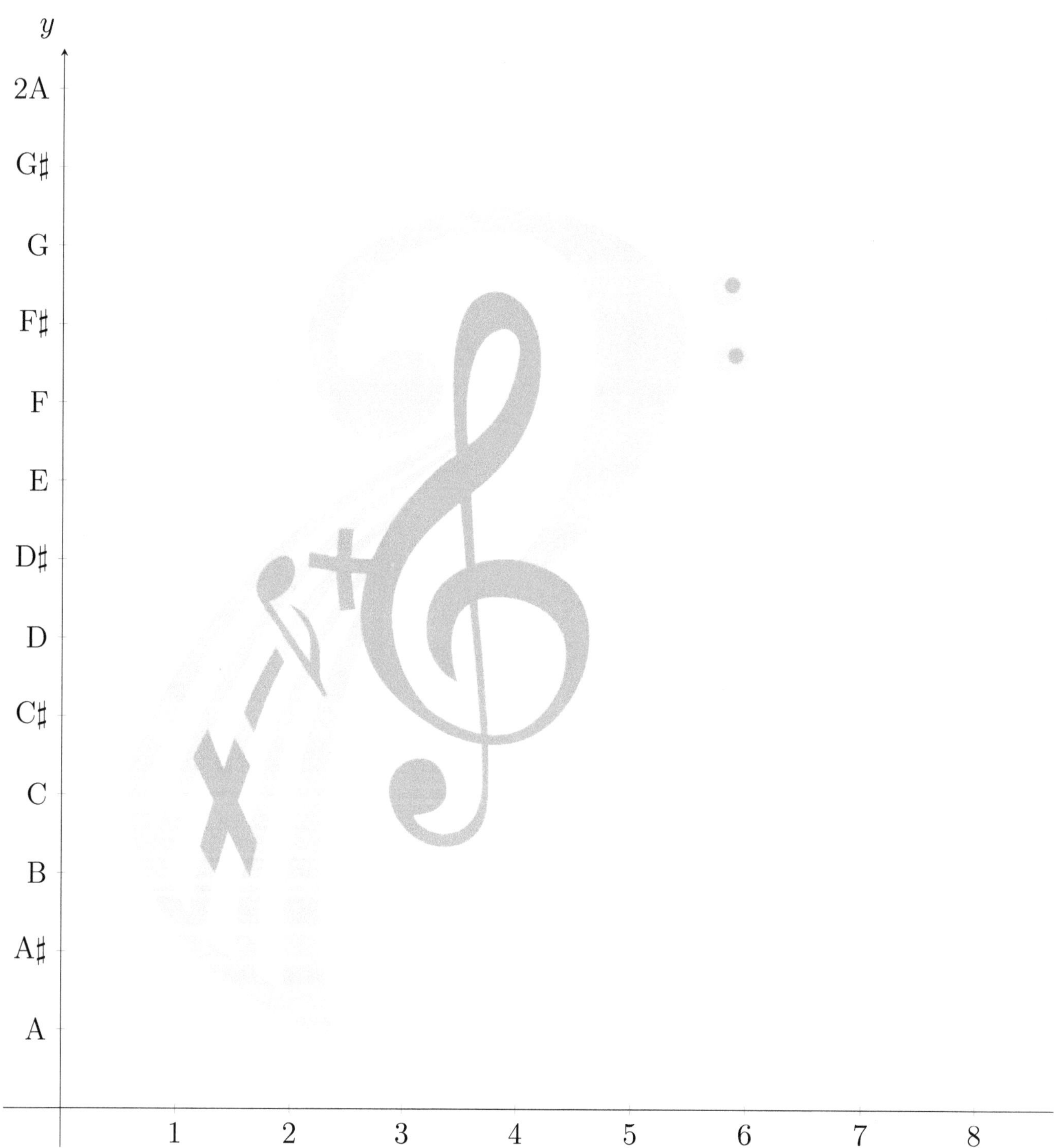

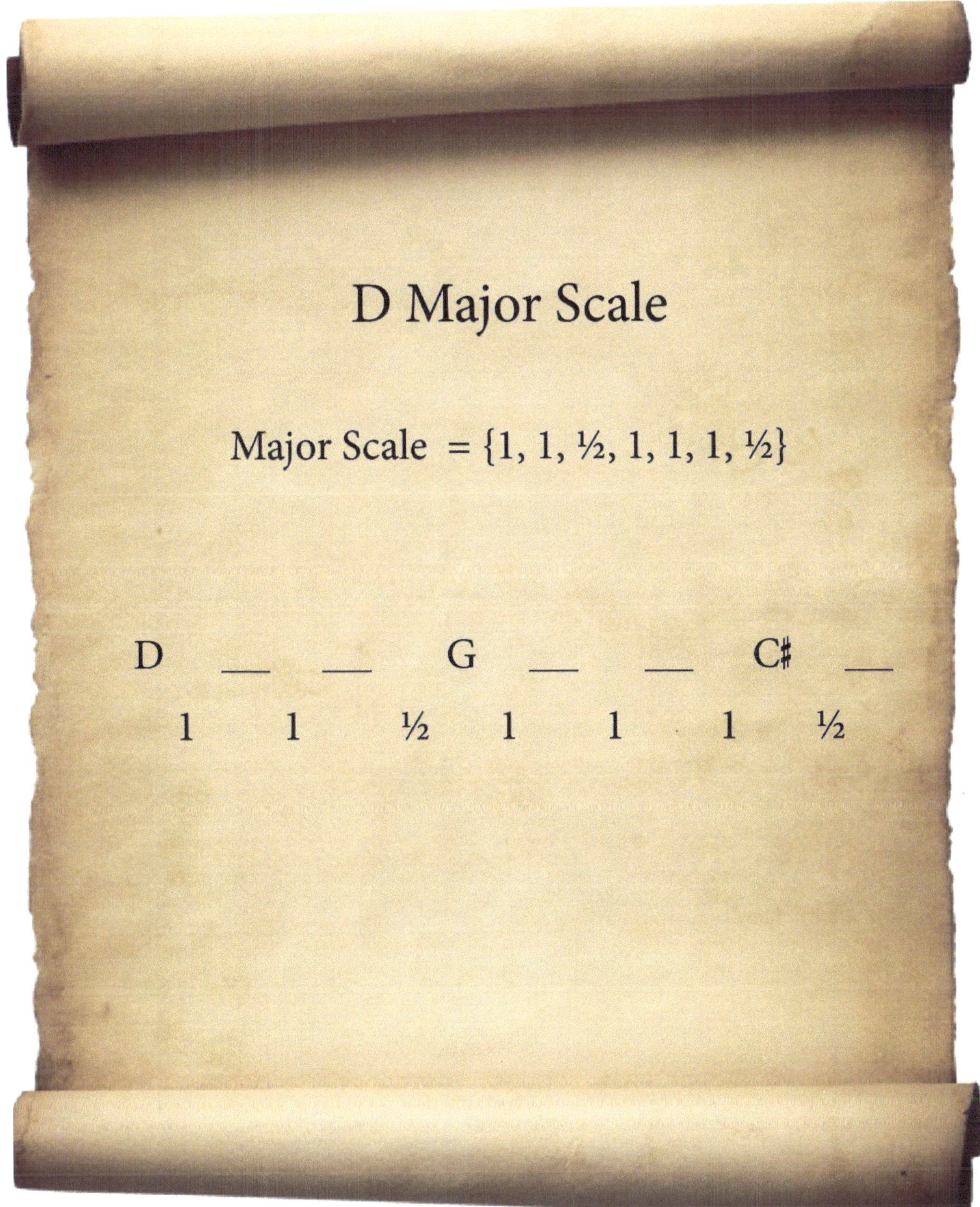
D Major Scale
Major Scale = {1, 1, ½, 1, 1, 1, ½}
D __ __ G __ __ C♯ __
1 1 ½ 1 1 1 ½

# D Major Scale

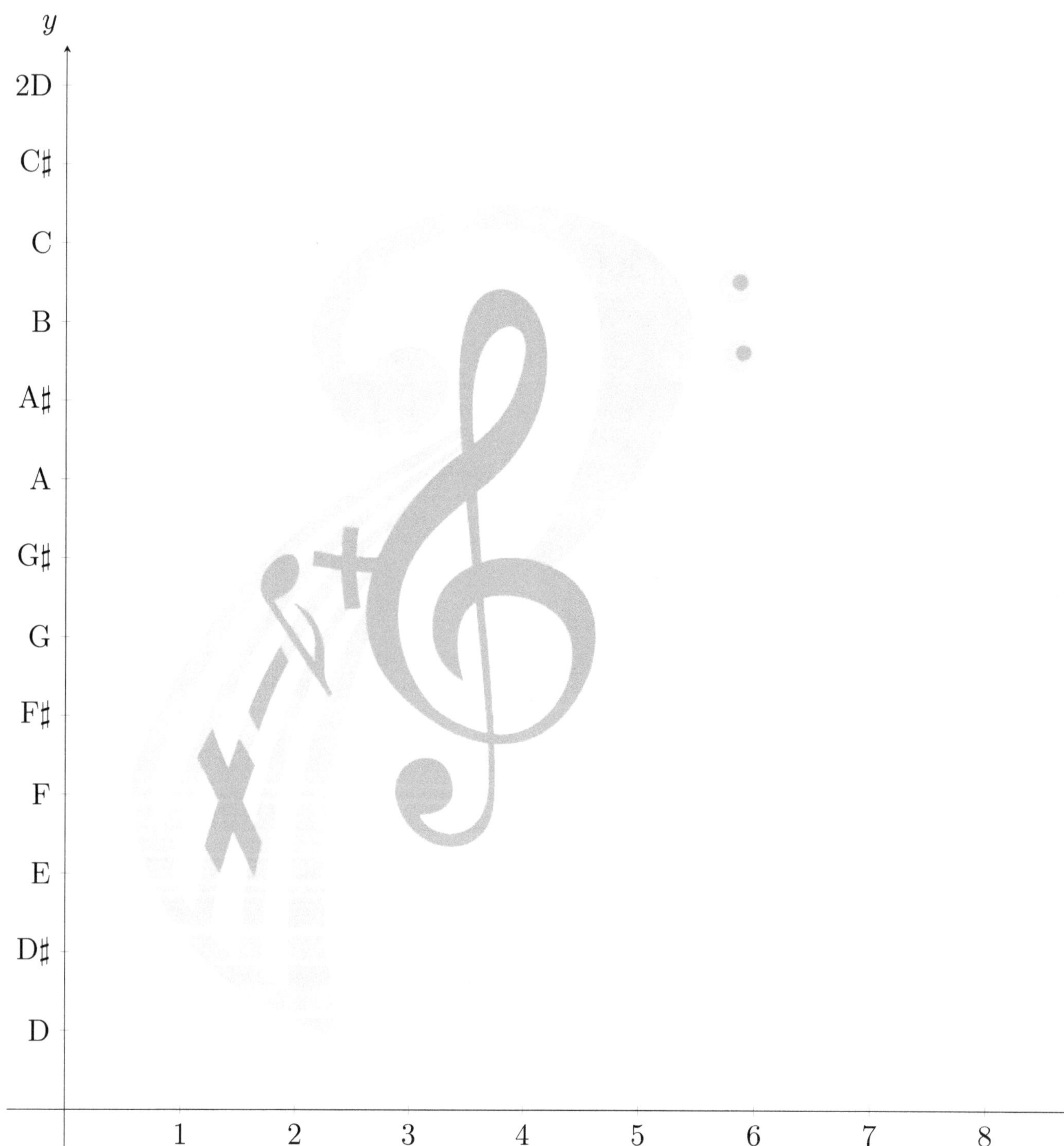

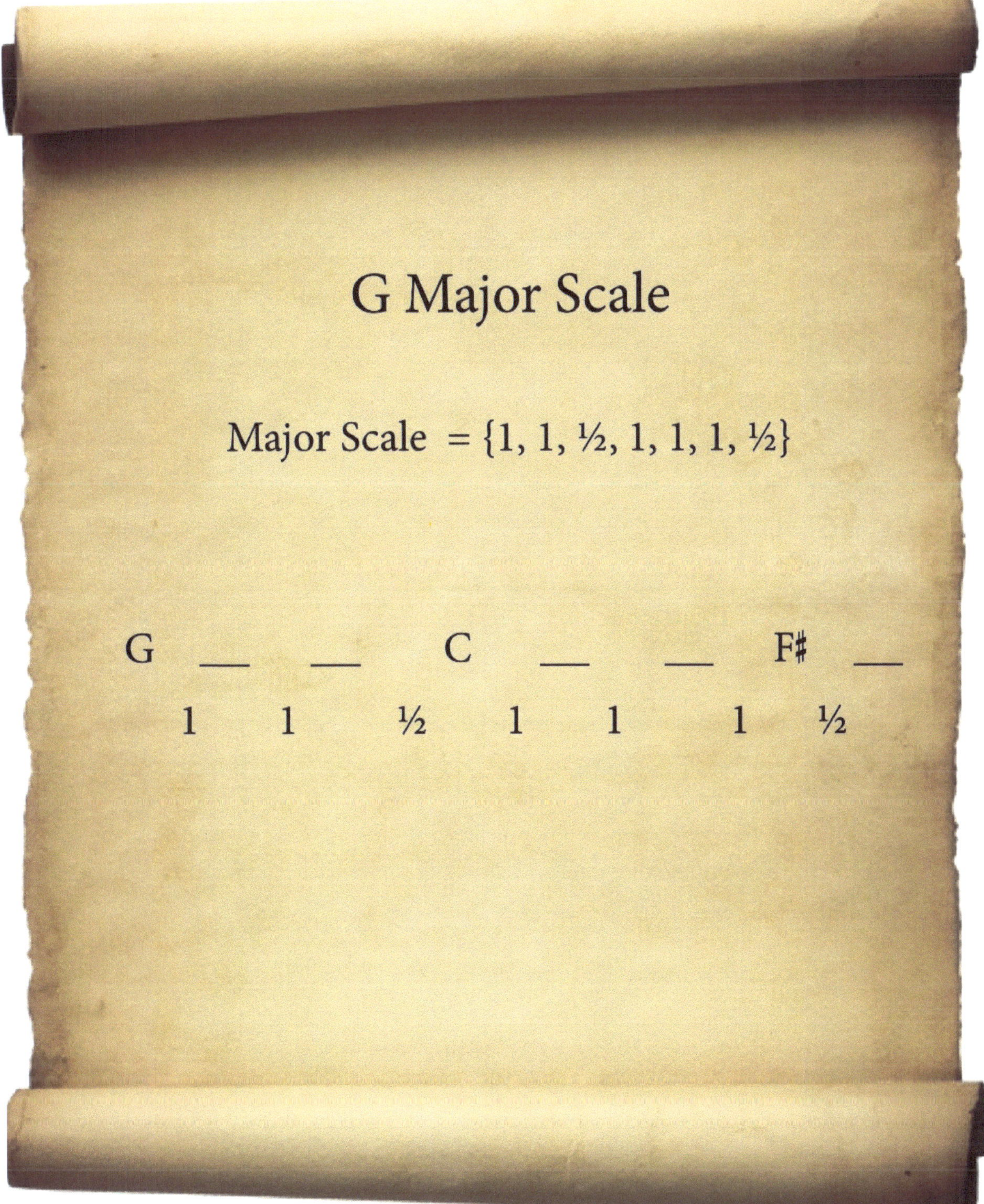
G Major Scale
Major Scale = {1, 1, ½, 1, 1, 1, ½}
G __ __ C __ __ F♯ __
1 1 ½ 1 1 1 ½

# G Major Scale

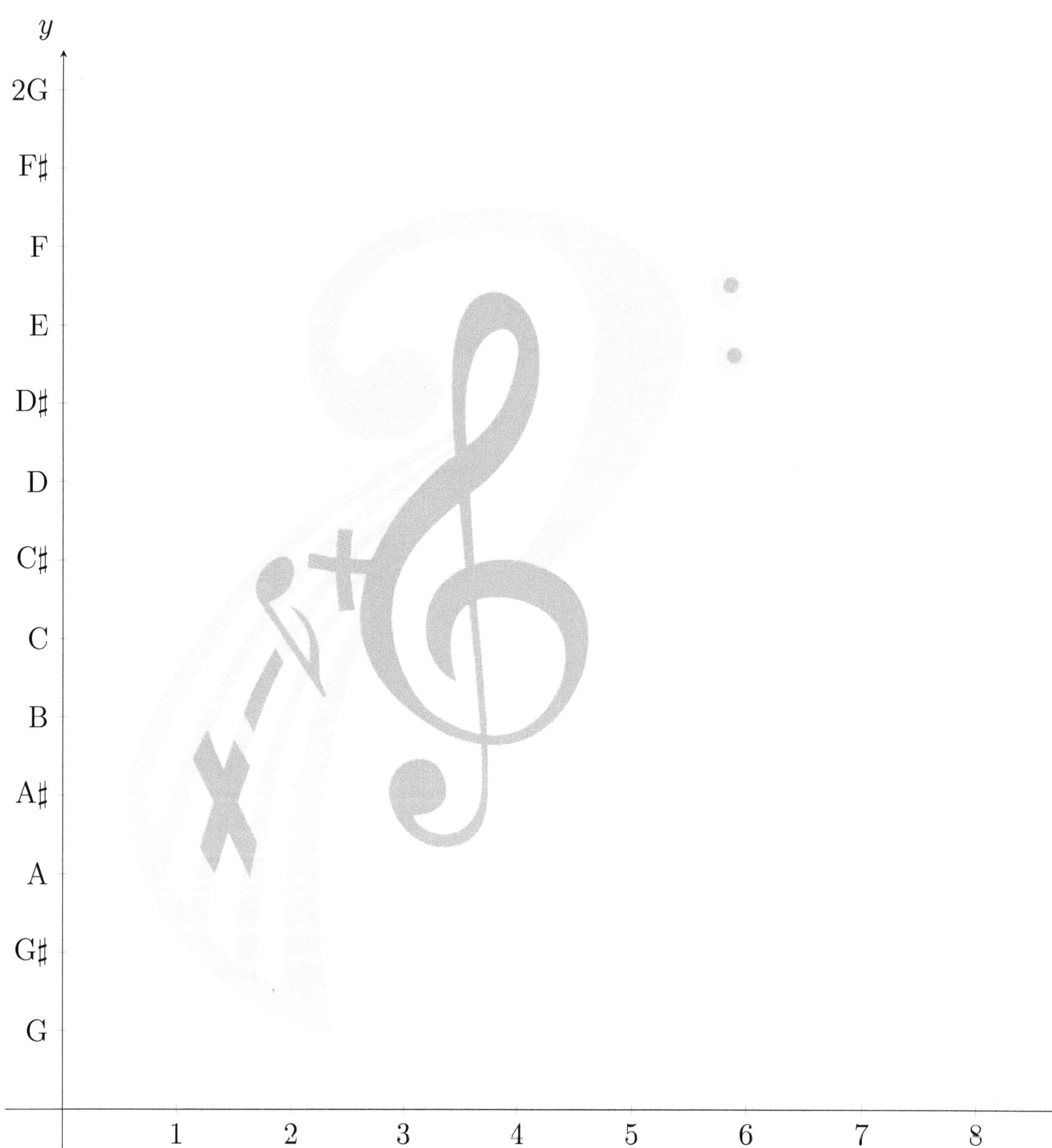

# 4 Warm Up Songs

Now that you have practiced graphing the major scales you will now graph the notes and melodies of popular songs.

# Rae Sremmurd - No Flex Zone

Please plot the following points in order to play the following song.

(1, A♭), (2, 2A♭), (3, E), (4, D♭)

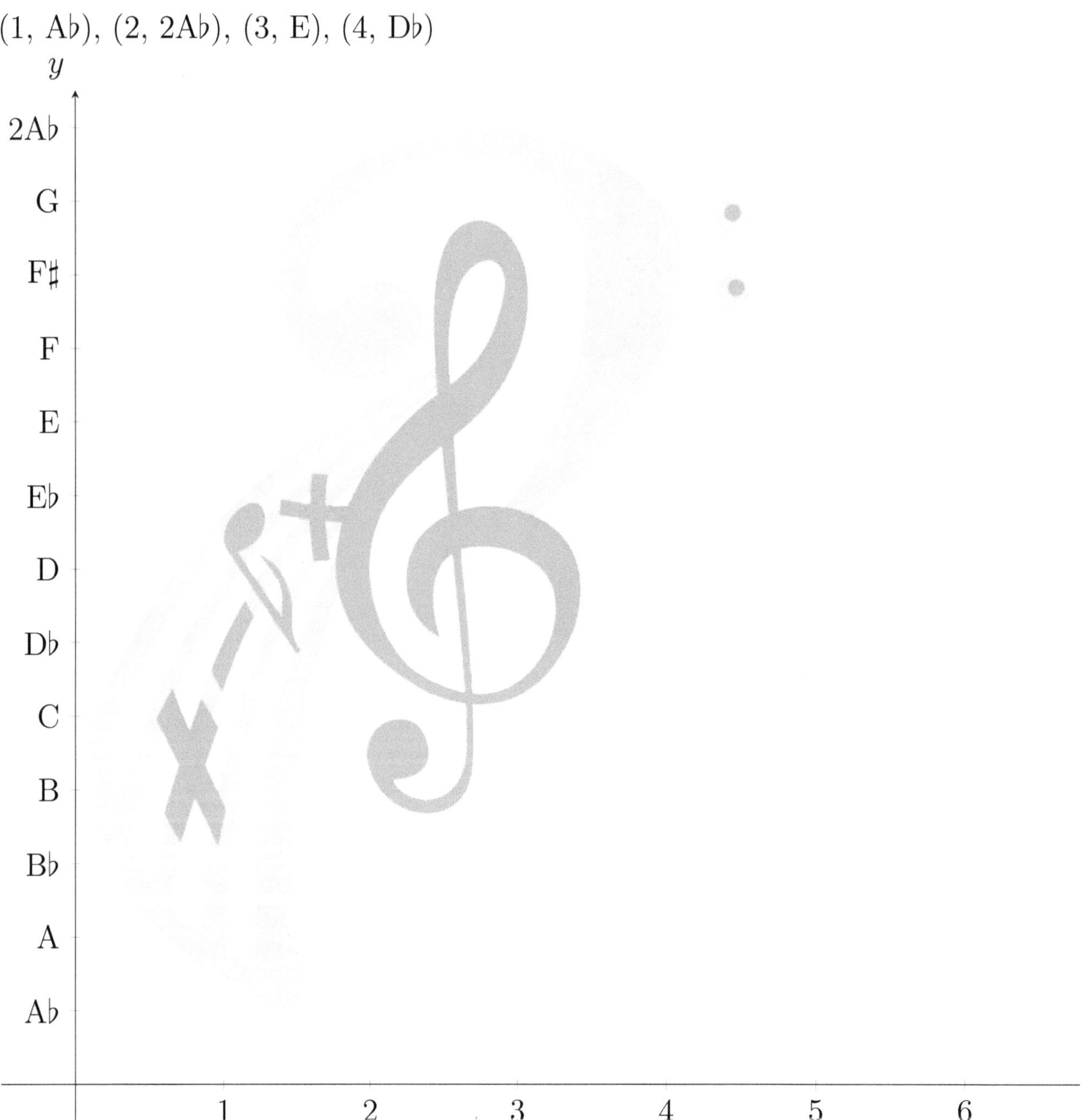

# Nicki Minaj ft. Beyonce - Feeling Myself

Please plot the following points in order to play the following song.

(1, C), (2, D), (3, F), (4, F), (5, A♭), (6, G)

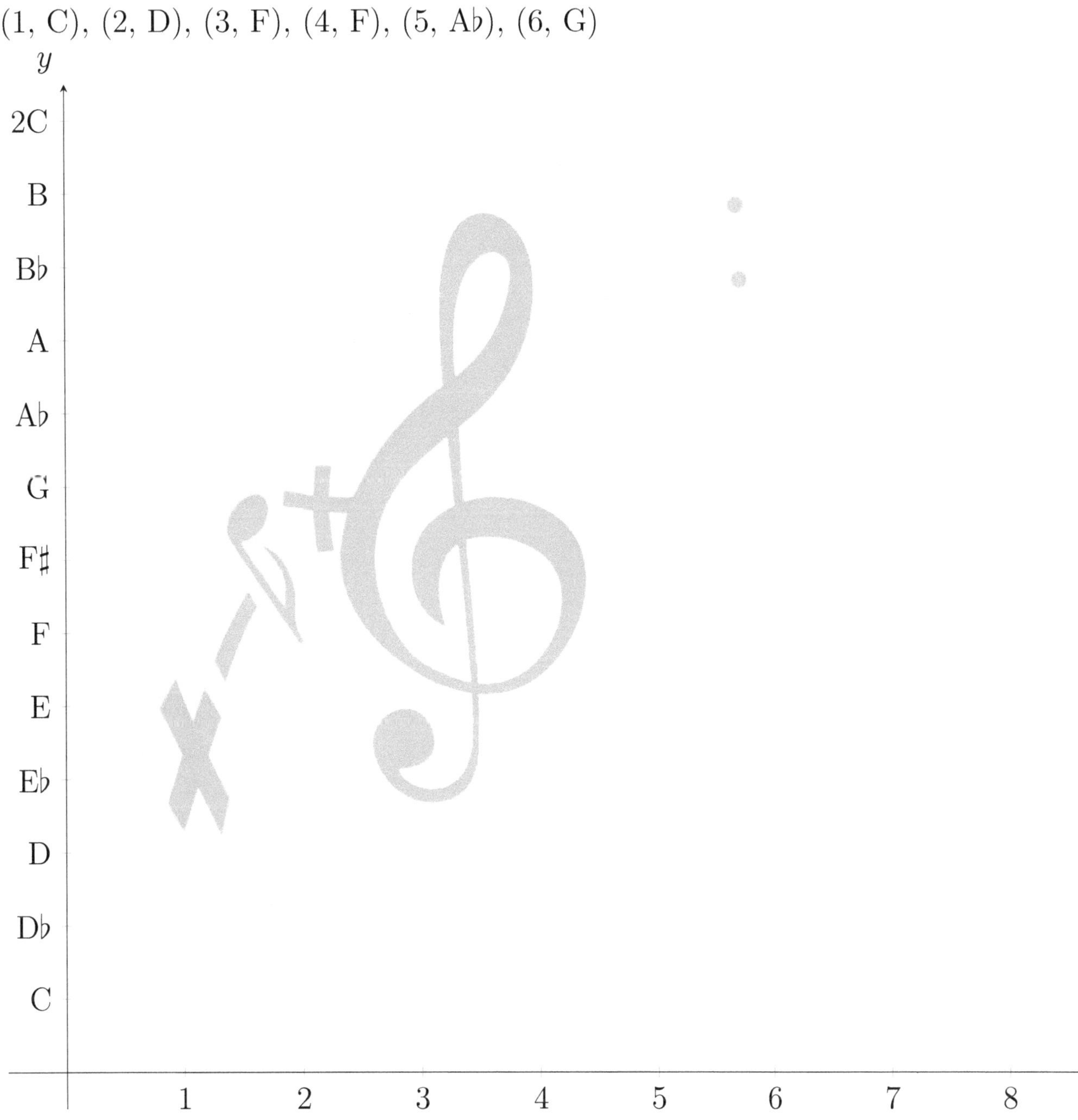

# 2 Chainz - I'm Different

Please plot the following points in order to play the following song.

(1, 2C), (2, B), (3, E), (4, A), (5, A), (6, F), (7, E)

# Nicki Minaj ft. Beyonce - Feeling Myself

Please plot the following points in order to play the following song.

(3, G),(7, D), (1, E), (4, A), (6, A), (2, F♯), (5, G), (8, 2E )

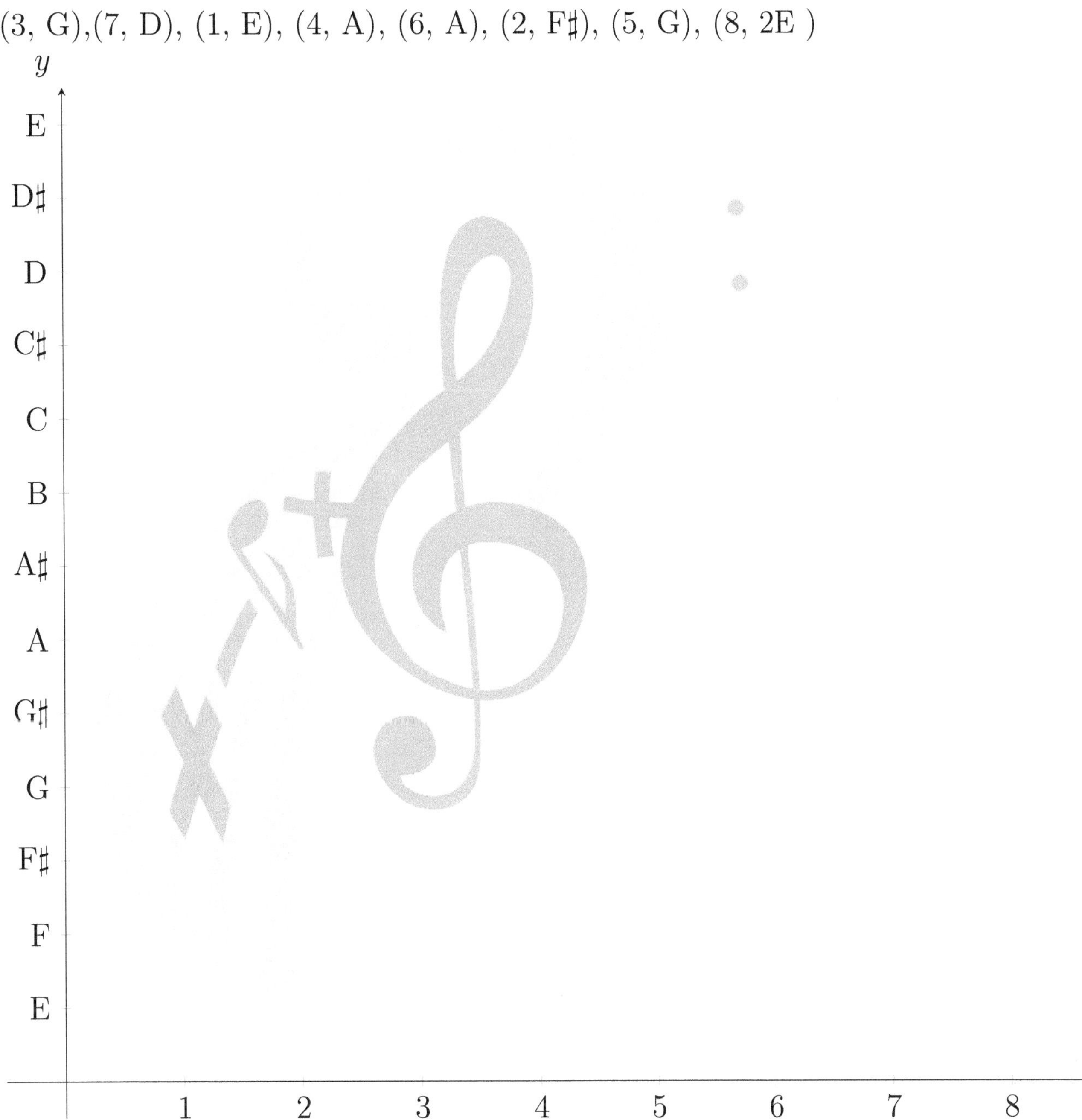

# Young Jeezy - R.I.P.

Please plot the following points in order to play the following song.

(4, F♯), (1, E), (3, G), (5, E♭) (2, E)

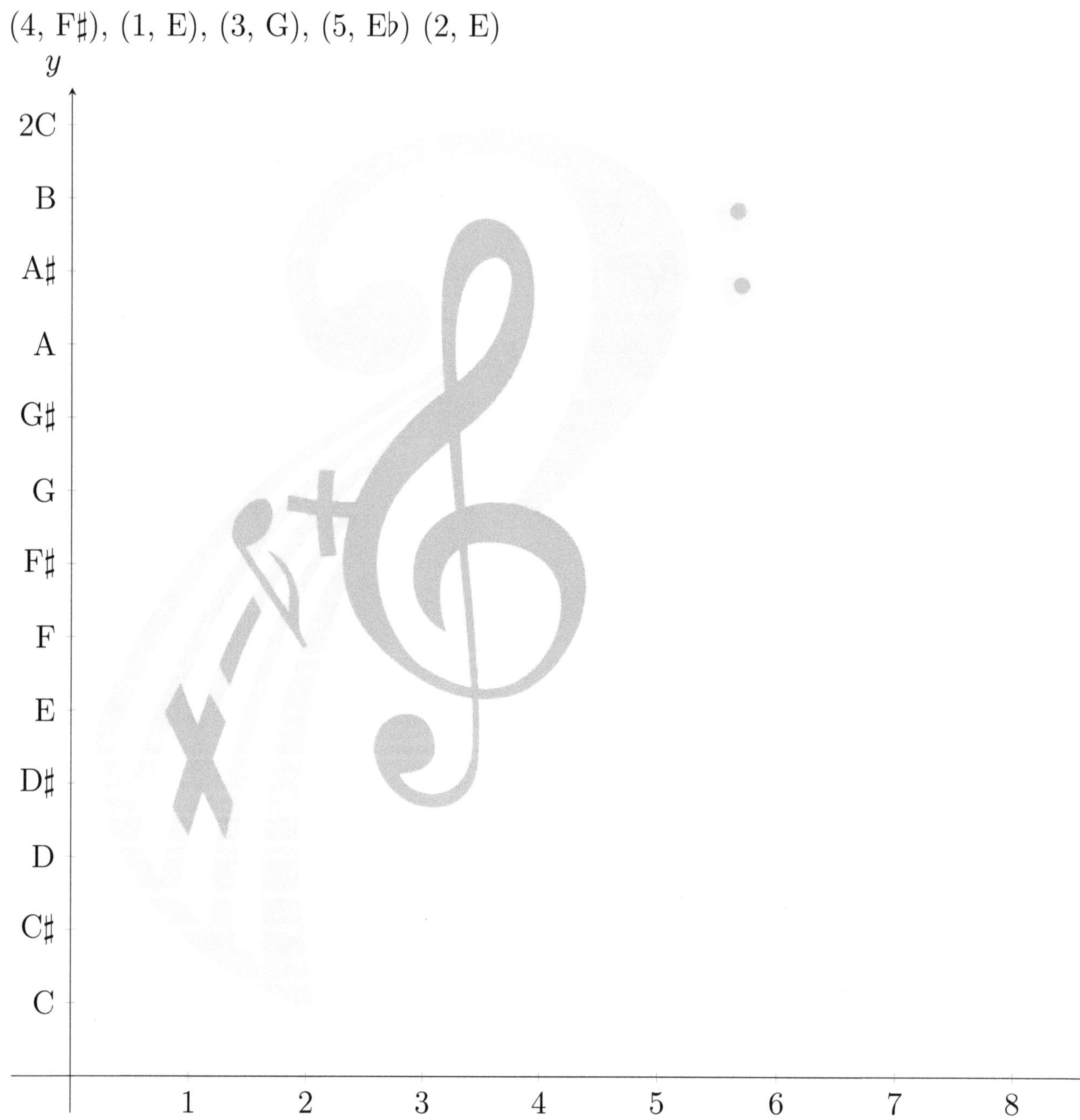

# Rae Sremmurd - Throw Sum Mo'

Please plot the following points in order to play the following song.

(5, D♭), (4, E♭), (6, E♭ ), (1, A♭), (3, E), (7, E) (8, F♯),(2, G♭),

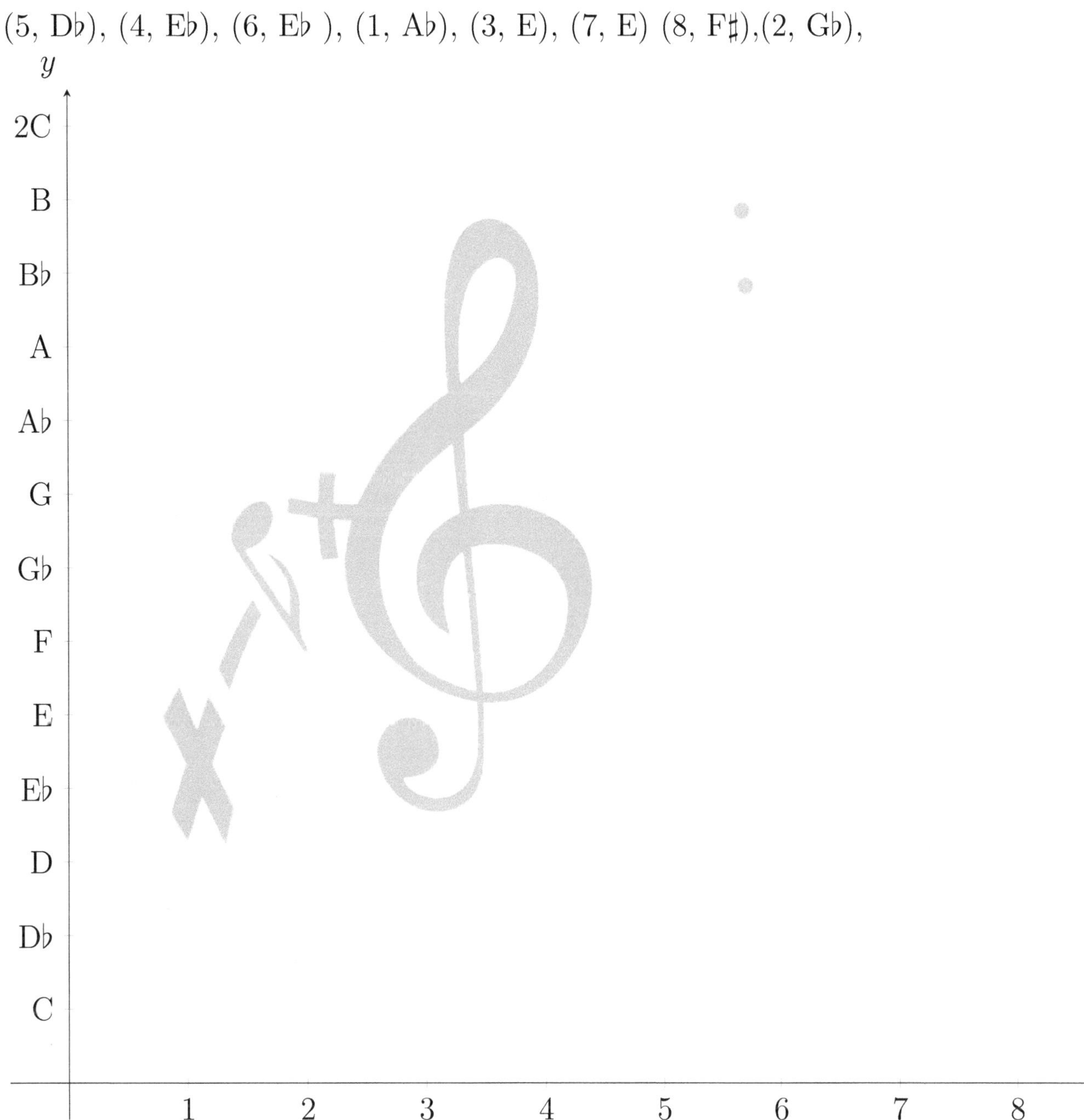

# Rae Sremmurd - No Type

Please plot the following points in order to play the following song.

(1, 2D), (1, B♭), (1, G ), (2, A), (2, F), (2, D)

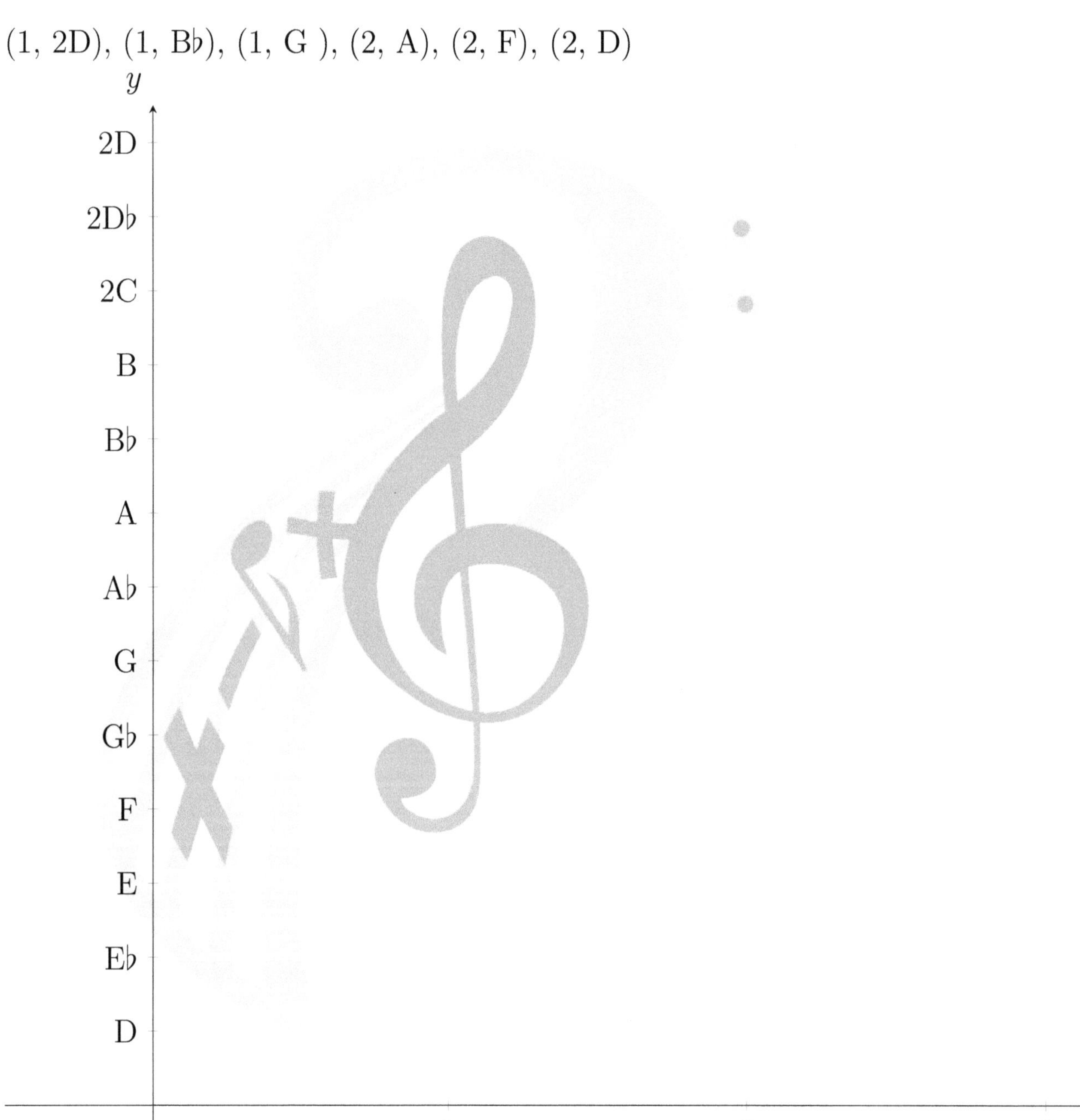

# Trinidad James - All Gold Everything

Please plot the following points in order to play the following song.

(1, A), (1, D), (2, B♭ ), (2, E♭), (3, G), (3, C)

## 5 Playing with Your Right And Left Hand

In this section you will begin to play songs with both hands. The way that you will distinguish what notes to play with each had will be by the musical notes plotted in specific quandrants. The musical notes to play with the right hand will be plotted in Quadrant 1. The notes to play with the left hand will be plotted in Quadrant 4.

In the following examples there will be many different lines to plot in one musical example and its very important that you remember how to play each song. The x values represent the order in which you play a note. So if you have to plot the points (1, C) (1, G) (1, E), all three of those coordinates would be played first and at the same time. What you will be playing is known as a musical chord which simply means you are playing at least three musical notes together at the same time.

Let's look at an example:

# Zedd ft. Foxes - Clarity

Please plot the following lines in order to play the featured song.

Line 1: (1, E♭), (2, E♭), (3, E♭), (4, E♭), (5, E♭)

Line 2: (1, B♭), (2, B♭), (3, C), (4, C), (5, C)

Line 3: (1, A♭), (2, G), (3, A♭), (4, A♭) (5, A♭)

Line 4: (1, -B♭), (2, -E♭), (3, -F), (4, -C), (5, -C♯)

# Zedd ft. Foxes - Clarity

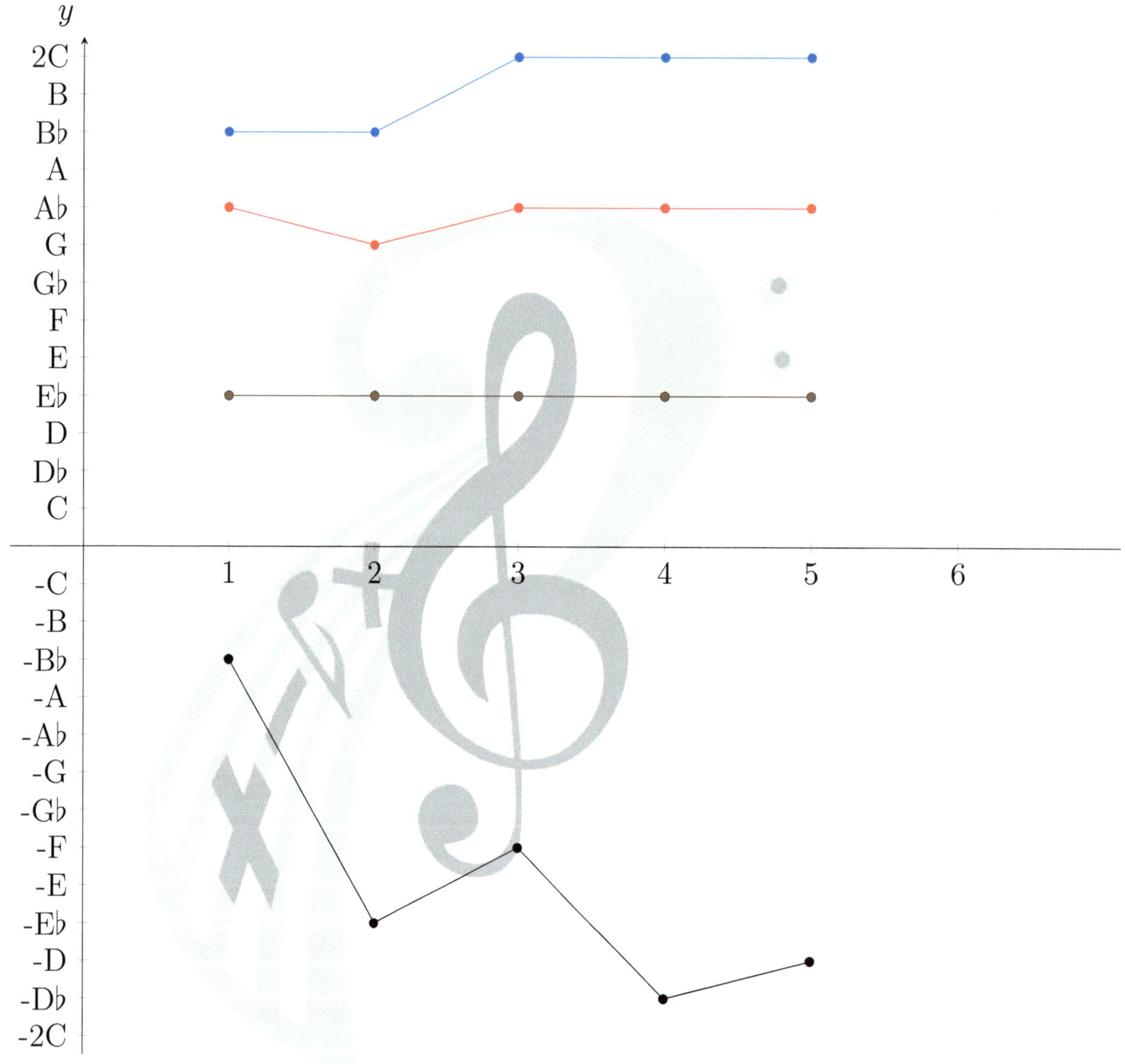

# Zedd ft. Foxes - Clarity (Solution)

Right Hand: Chord 1: B♭, A♭, E♭

Left Hand: B♭

Right Hand: Chord 2: B♭, G, E♭

Left Hand: E♭

Right Hand: Chord 3: C, A♭, E♭

Left Hand: B♭

Right Hand: Chord 4: C, A♭, E♭

Left Hand: B♭

Right Hand: Chord 5: C, A♭, E♭

Left Hand: B♭

# Justin Timberlake - Mirrors

Please plot the following lines in order to play the featured song.

Line 1: (1, E♭), (2, D), (3, C),

Line 2: (1, B♭), (2, B♭), (3, A♭),

Line 3: (1, G), (2, F), (3, E♭),

Line 4: (1, -E♭), (2, -B♭), (3, -A♭),

## Justin Timberlake - Mirrors

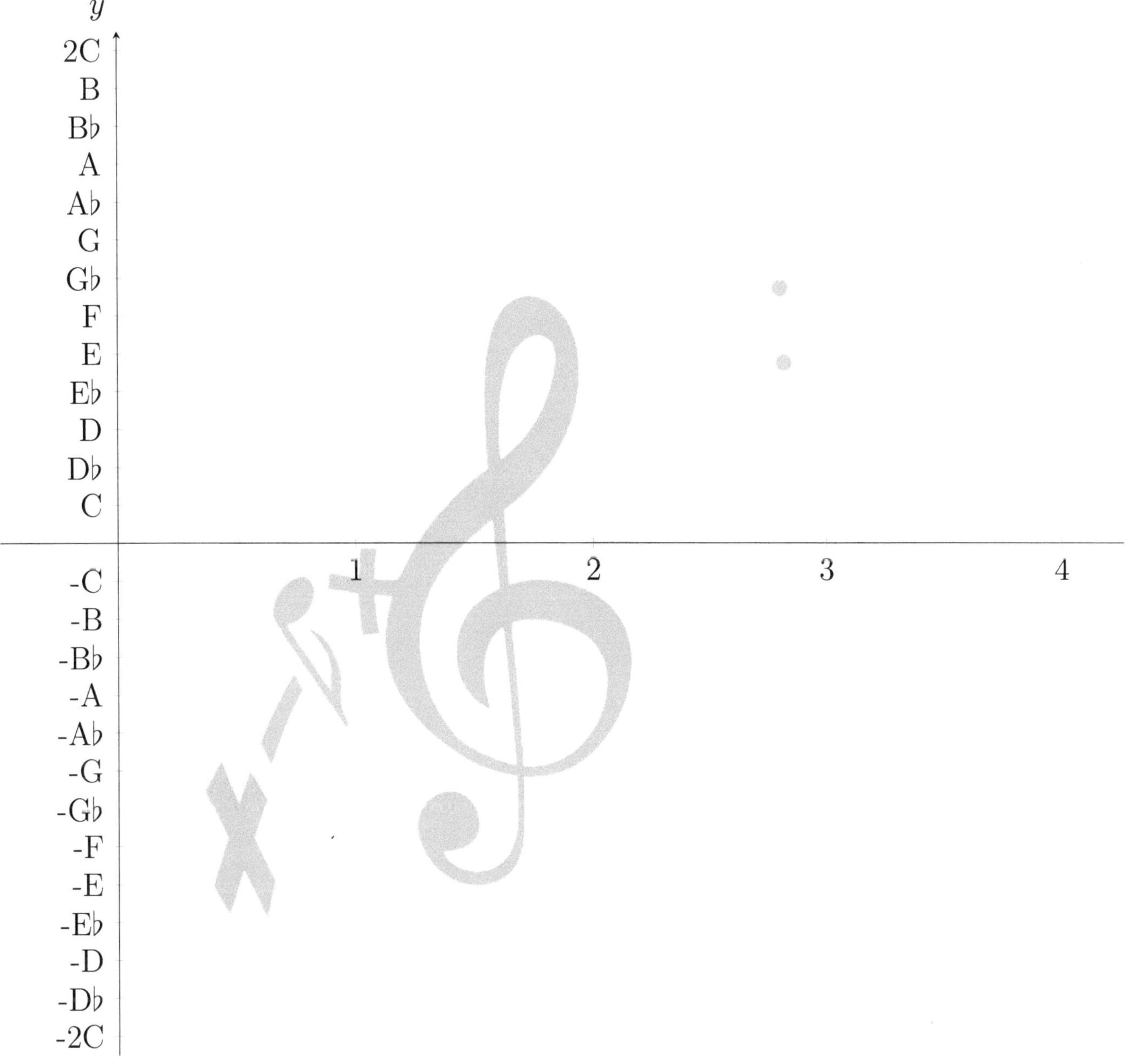

# Rihanna - We Found Love

Please plot the following lines in order to play the featured song.

Line 1: (1, B), (2, D), (3, B),(4, D)

Line 2: (1, G), (2, B), (3, G), (4, B)

Line 3: (1, E), (2, G), (3, E), (4, G)

Line 4: (1, -E), (2, -C), (3, -G), (4, -G♭

# Rihanna - We Found Love

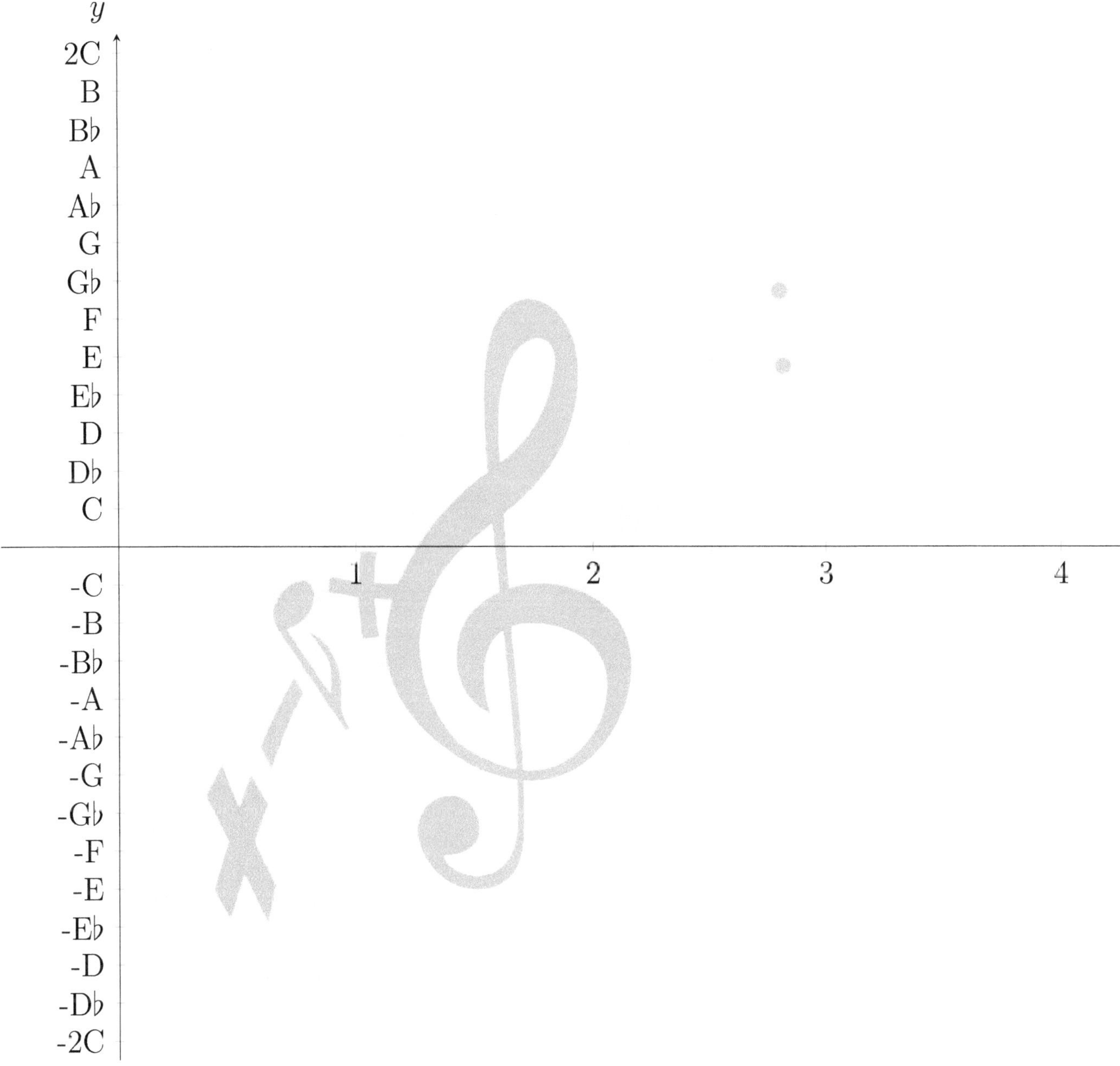

# Cash Out - Cashin Out

Please plot the following lines in order to play the featured song.

Line 1: (1, A♭), (2, E), (3, D♭),(4, A), (5, A♭), (6, G♭)

Line 2: (1, -A♭), (3, -A♭), (4, -A), (6, -A)

Line 3: (1, -E), (3, -E), (4, -F♯), (6, -F♯)

Line 4: (1, -D♭), (3, -*D*♭), (4, -C♯), (6, -C♯

# Cash Out - Cashin Out

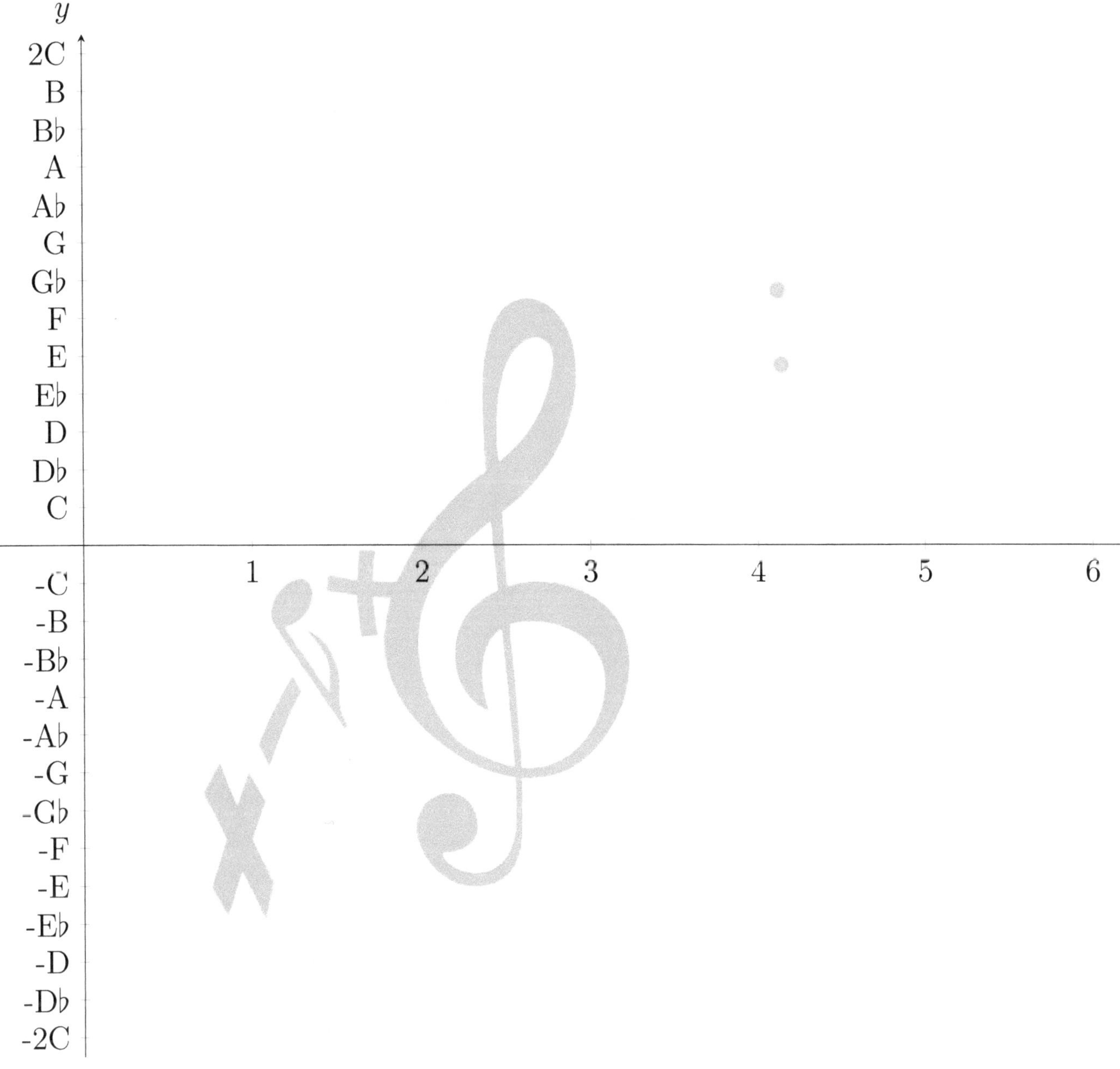

# Ace Hood - Hustle Hard

Please plot the following lines in order to play the featured song.

Line 1: (1, G), (2, F♯), (3, B),(4, G), (5, F♯), (6, B)

Point 1: (1, -G♭)

Point 2: (1, -D)

Point 3: (1, -B)

# Ace Hood - Hustle Hard

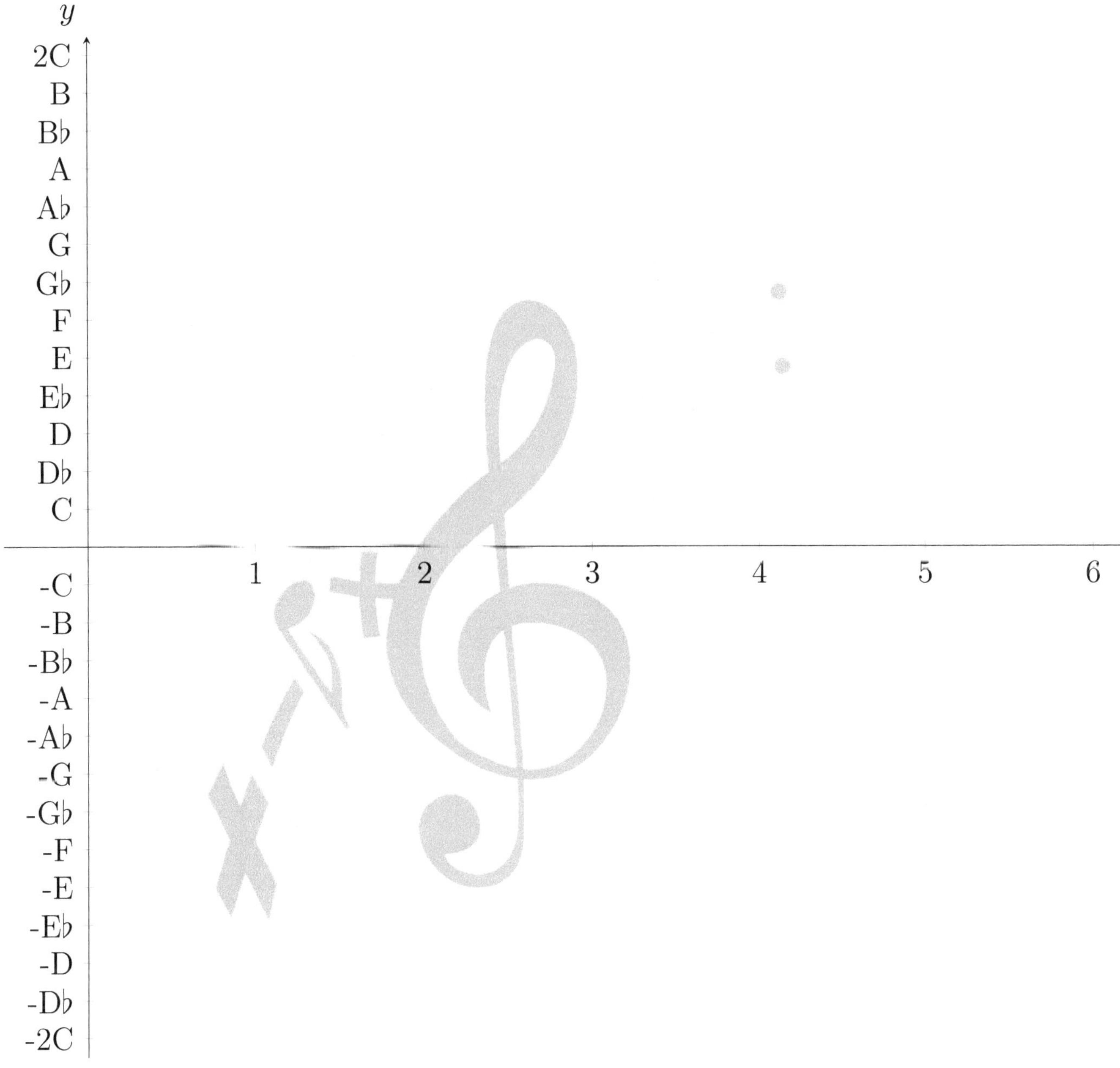

# Kendrick Lamar - Poetic Justice

Please plot the following lines in order to play the featured song.

Line 1: (1, A♭), (2, A♭), (3, A♭),(4, F♯), (5, F♯)

Line 2: (1, E), (2, E), (3, E), (4, E♭), (5, E♭)

Line 3: (1, B), (2, B), (3, B), (4, C), (5, C)

Line 4 : (1, -D♭), (2, -A♭), (3, -G♭), (4, -A♭), (5, -A♭)

# Kendrick Lamar - Poetic Justice

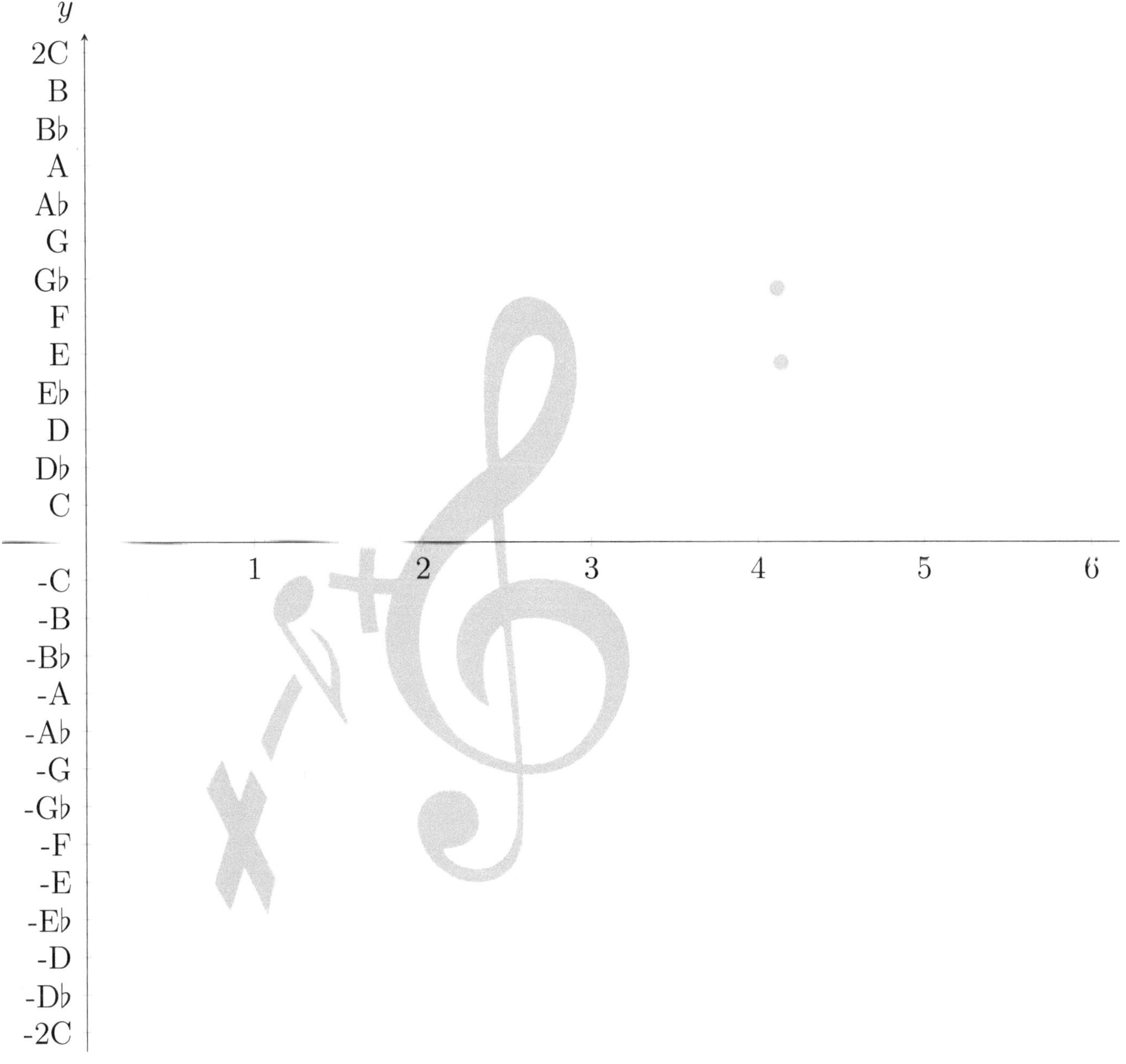

# Rihanna - Nobody Business

Please plot the following lines in order to play the featured song.

Line 1: (1, A♭), (2, F♯), (3, A♭),

Line 2: (1, E), (2, E♭), (3, F),

Line 3: (1, B), (2, B), (3, D♭)

Line 4 : (1, -A), (2, -D♭), (3, -D♭)

# Rihanna - Nobody Business

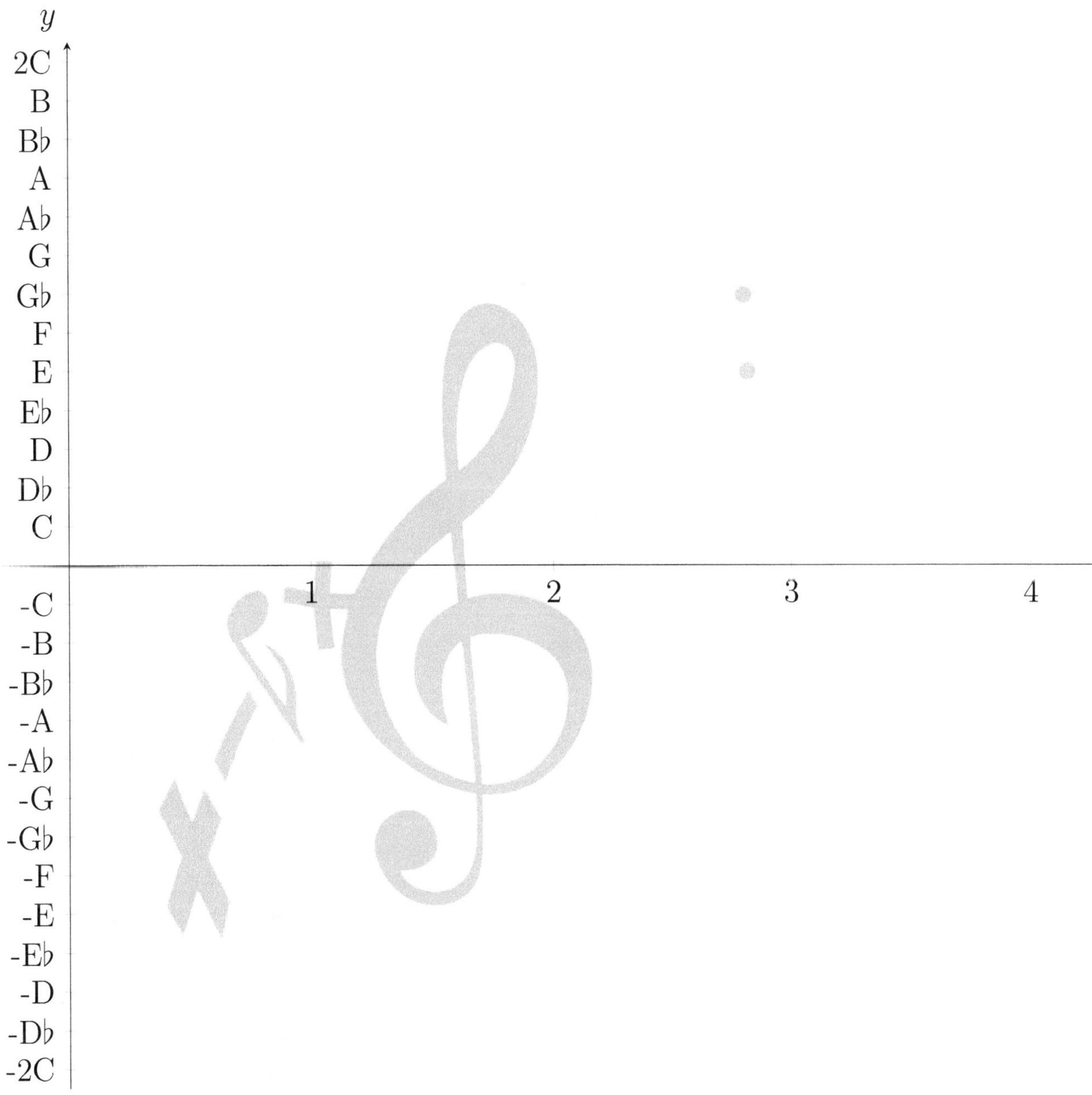

# Janelle Monae - Q.U.E.E.N.

Please plot the following lines in order to play the featured song.

Line 1: (1, A), (2, G♯), (3, F♯)

Line 2: (1, F♯), (2, E), (3, D)

Line 3: (1, D), (2, B), (3, B)

Line 4 : (1, -B), (2, -G♭), (3, -D)

# Janelle Monae - Q.U.E.E.N.

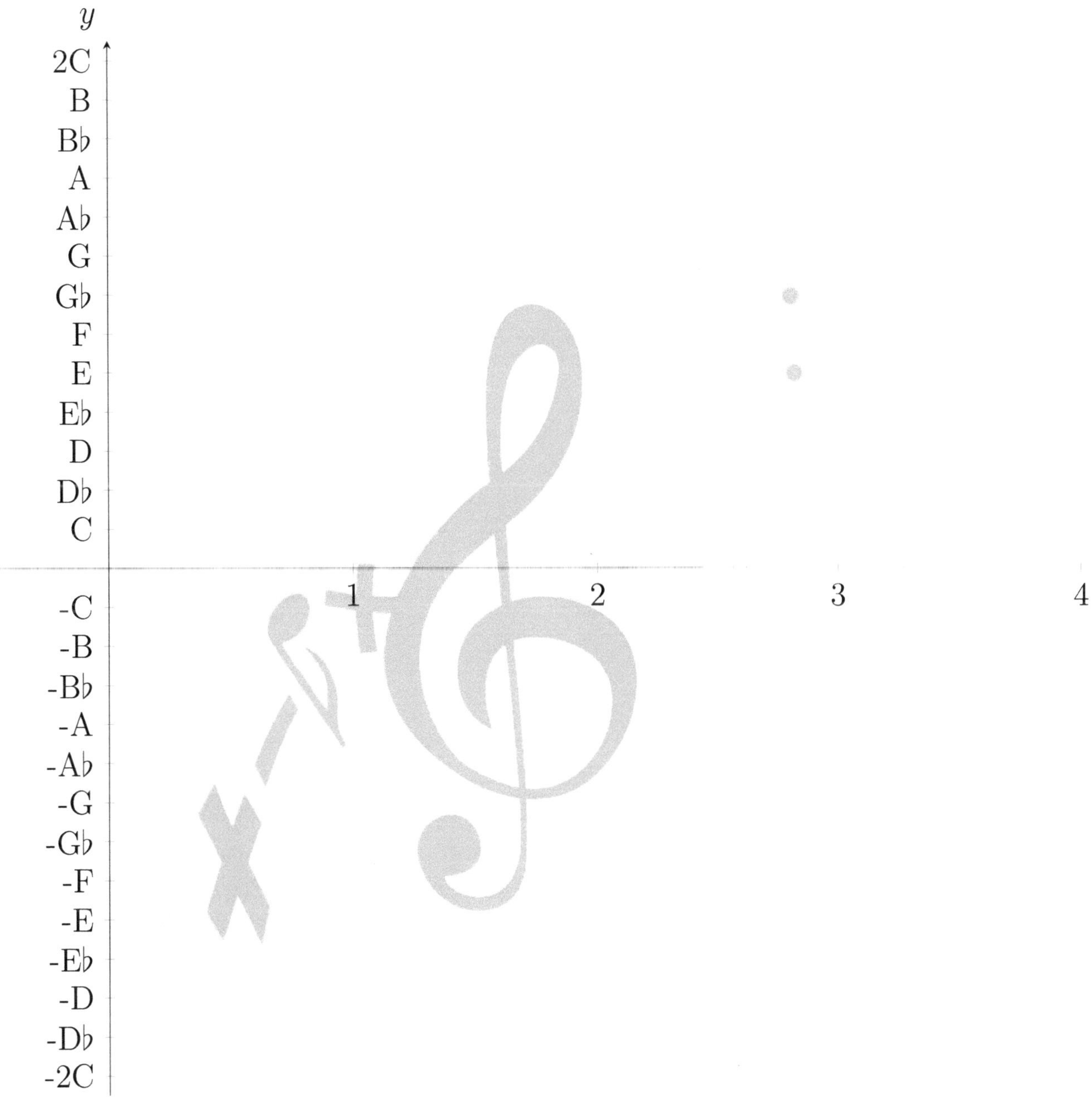

# French Montana - Marble Floors

Please plot the following lines in order to play the featured song.

Line 1: (1, E♭), (2, F♯), (3, B♭),(4, B)

Line 1: (1, -E♭), (2, -E♭), (3, -E♭), (4, -E♭)

Line 2: (1, -B♭), (2, -B♭), (3, -B♭), (4, -B♭)

Line 3: (1, -G♭), (2, -G♭), (3, -G♭), (4, -G♭)

# French Montana - Marble Floors

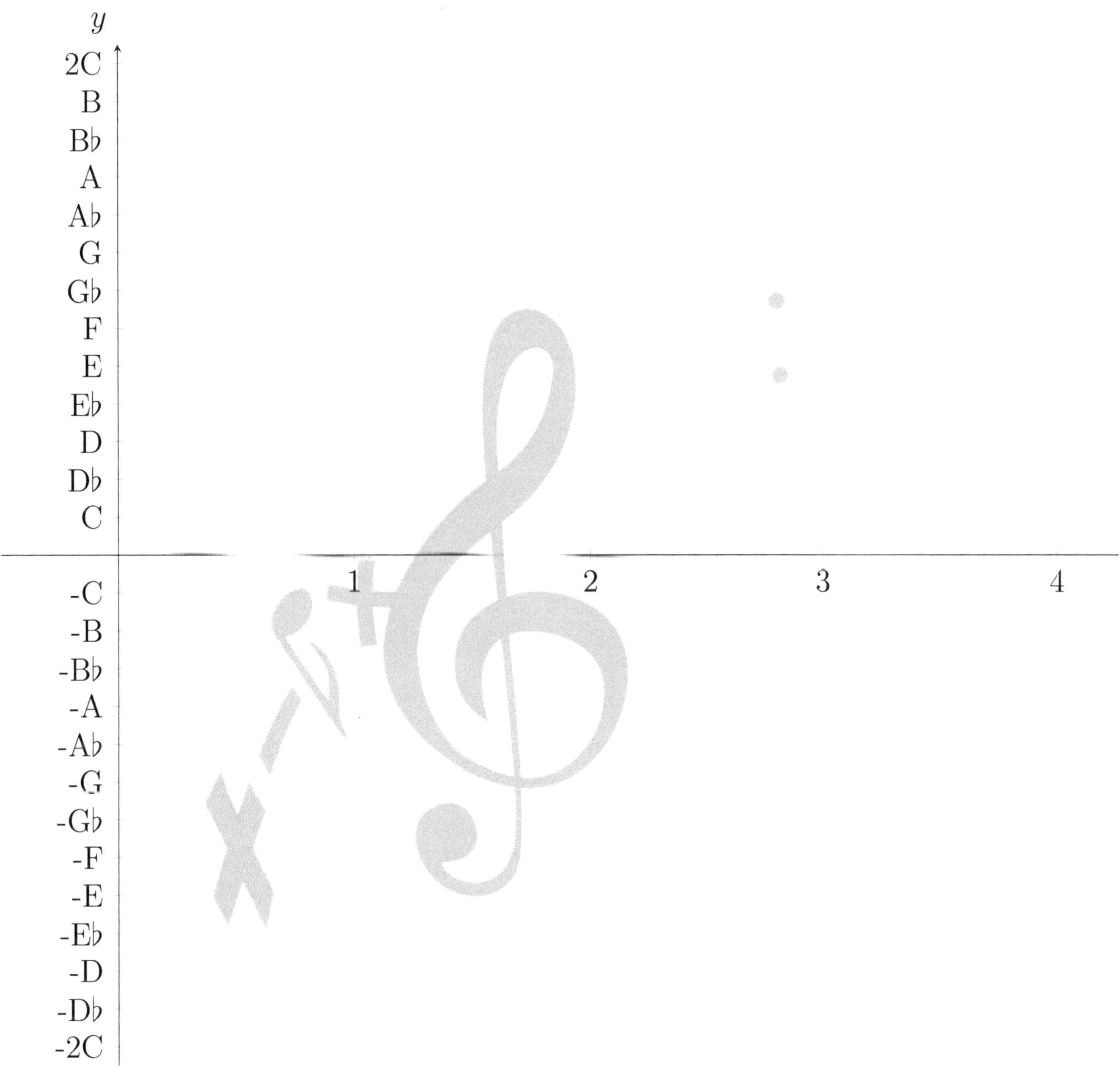

# Black Eyed Peas - Gotta Feeling

Please plot the following lines in order to play the featured song.

Line 1: (1, G), (2, G), (3, G), (4, G)

Line 2: (1, D), (2, E), (3, D), (4, E)

Line 3: (1, B), (2, C), (3, B), (4, C)

Line 4 : (1, -G), (2, -C), (3, -E), (4, -C)

# Black Eyed Peas - Gotta Feeling

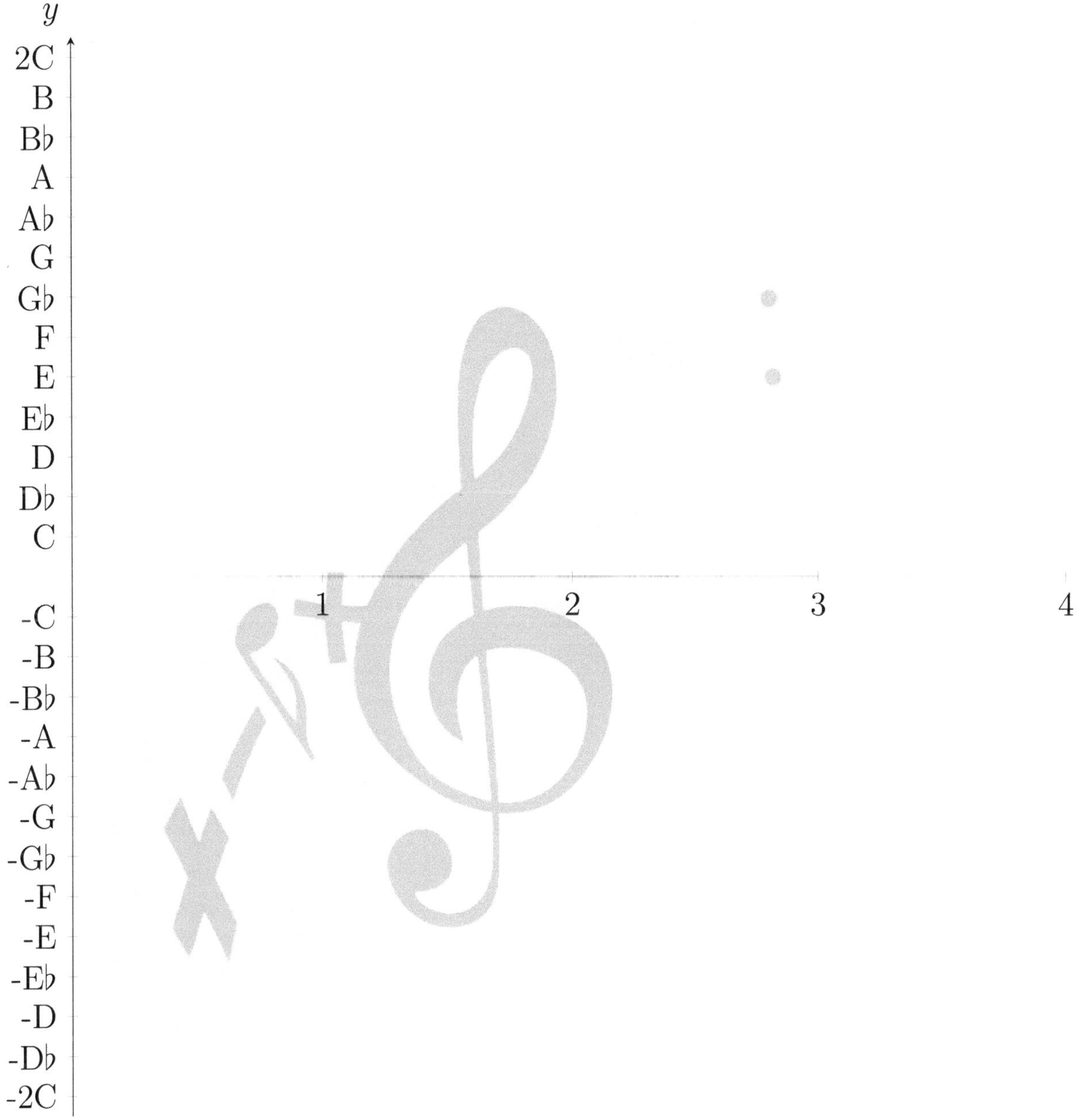

# Katy Perry - Firework

Please plot the following lines in order to play the featured song.

Line 1: (1, C), (2, C♯), (3, C), (4, C)

Line 2: (1, A♭), (2, G♯), (3, A♭), (4, A♭)

Line 3: (3, F), (4, F)

Line 4 : (1, -A♭), (2, -B♭), (3, -F), (4, -D♭)

# Katy Perry - Firework

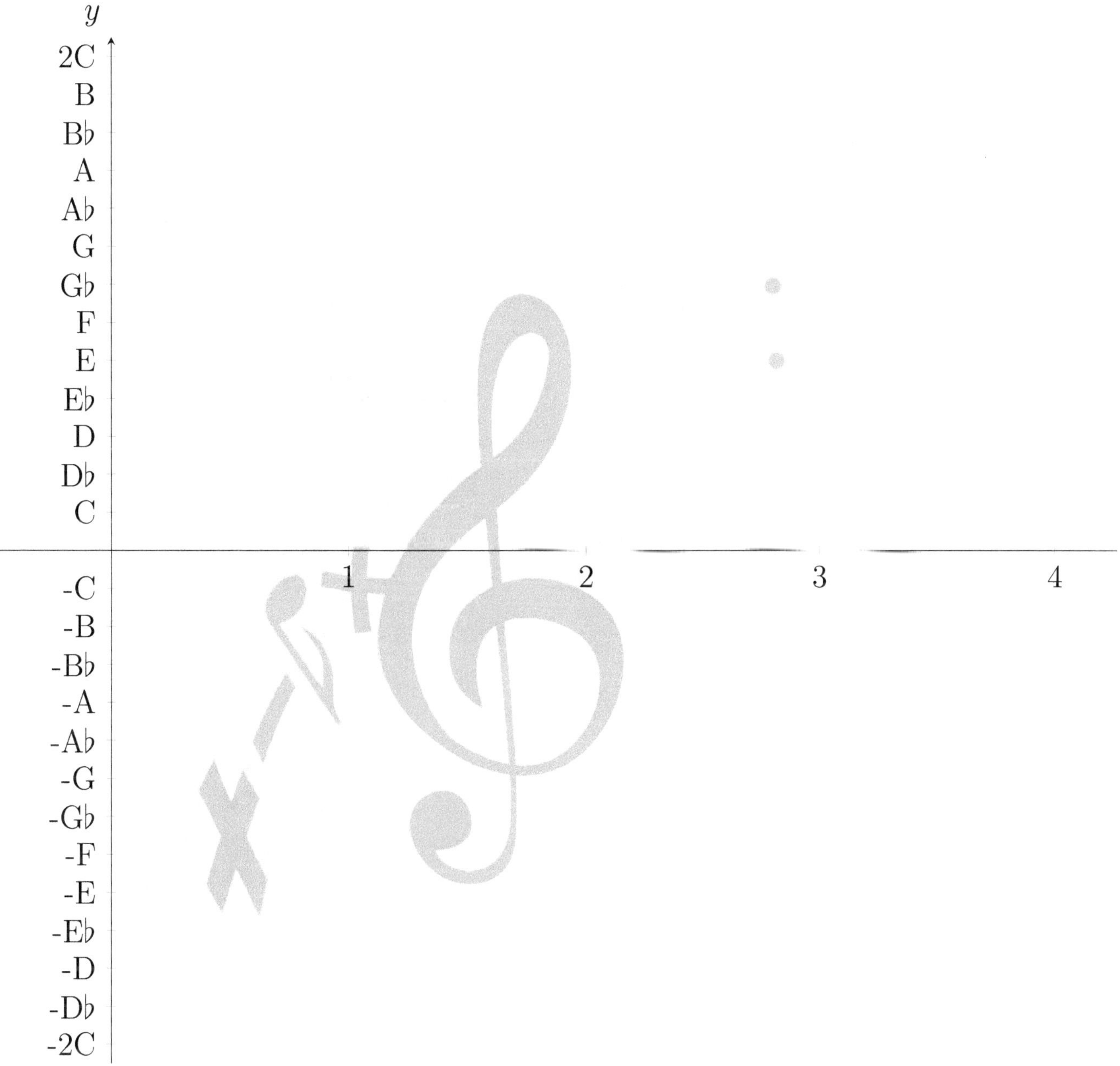

# Lil Wayne - How to Love

Please plot the following lines in order to play the featured song.

Line 1: (1, E♭), (2, E♭), (3, E♭), (4, E)

Line 2: (1, B), (2, B♭), (3, B), (4, B)

Line 3: (1, F♯), (2, F♯), (3, F♯), (4, G)

Line 4 : (1, -B), (2, -B), (3, -E), (4, -E)

# Lil Wayne - How to Love

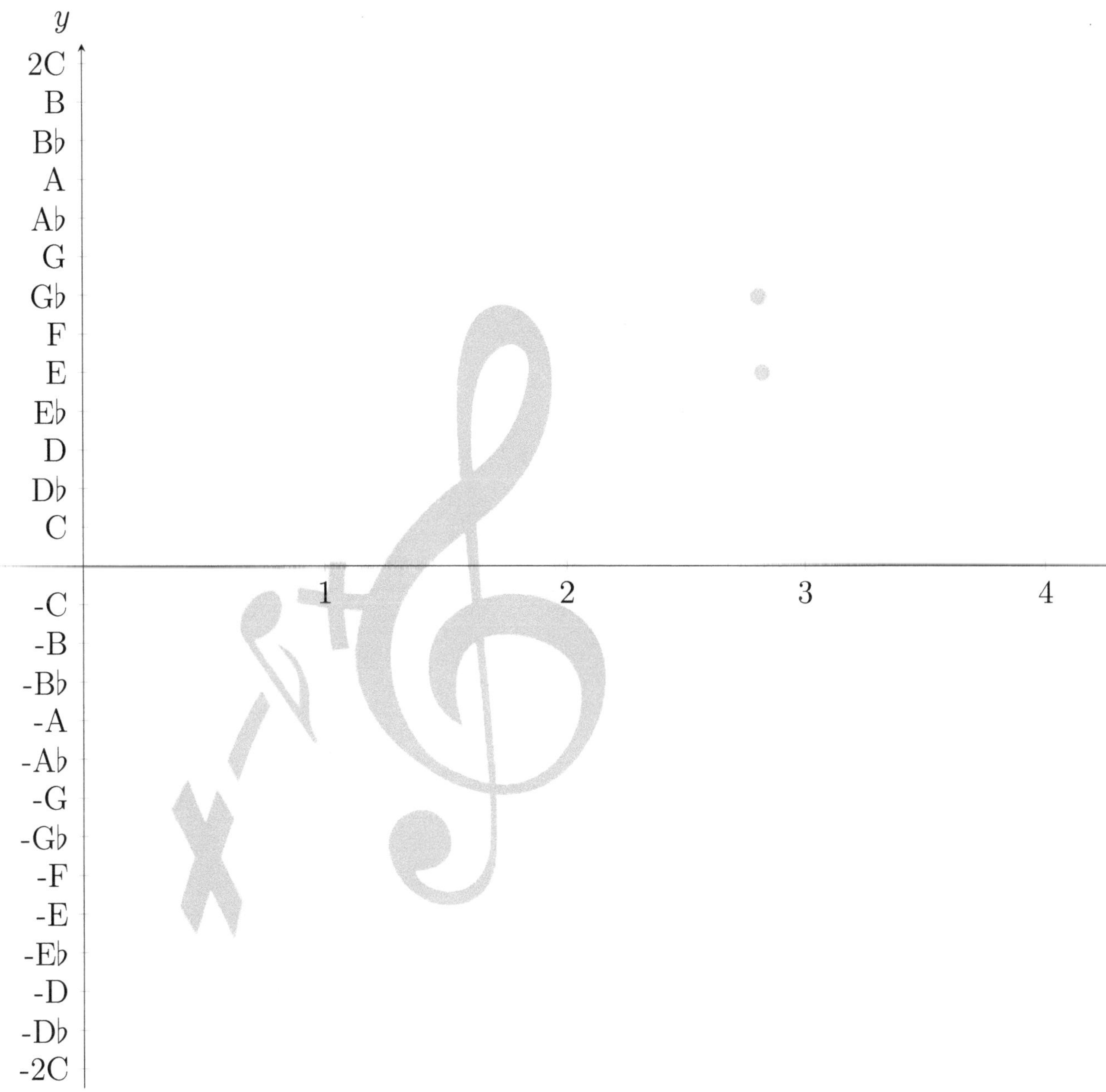

# Miley Cyrus - We Can't Stop

Please plot the following lines in order to play the featured song.

Line 1: (1, A♭), (2, G♭), (3, A♭), (4, A)

Line 2: (1, E), (2, E♭), (3, E), (4, E)

Line 3: (1, B), (2, B), (3, B), (4, D♭)

Line 4 : (1, -E), (2, -A♭), (3, -D♭), (4, -A)

# Miley Cyrus - We Can't Stop

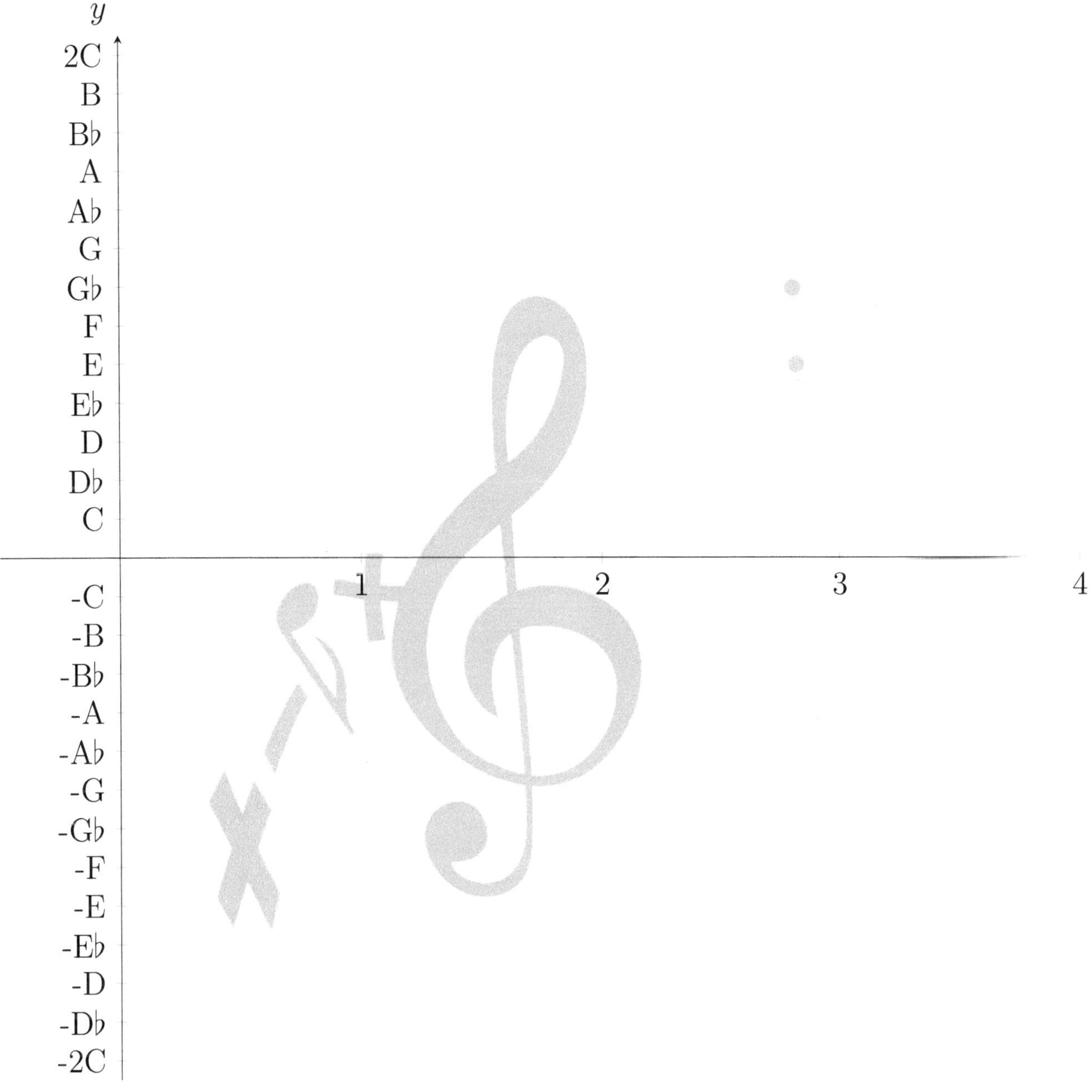

# Rihanna - Diamonds

Please plot the following lines in order to play the featured song.

Line 1: (1, D), (2, G♭), (3, E)

Line 2: (1, B), (2, D), (3, D♭)

Line 3: (1, G), (2, A), (3, A)

Line 4 : (1, -G), (2, -B), (3, -A)

# Rihanna - Diamonds

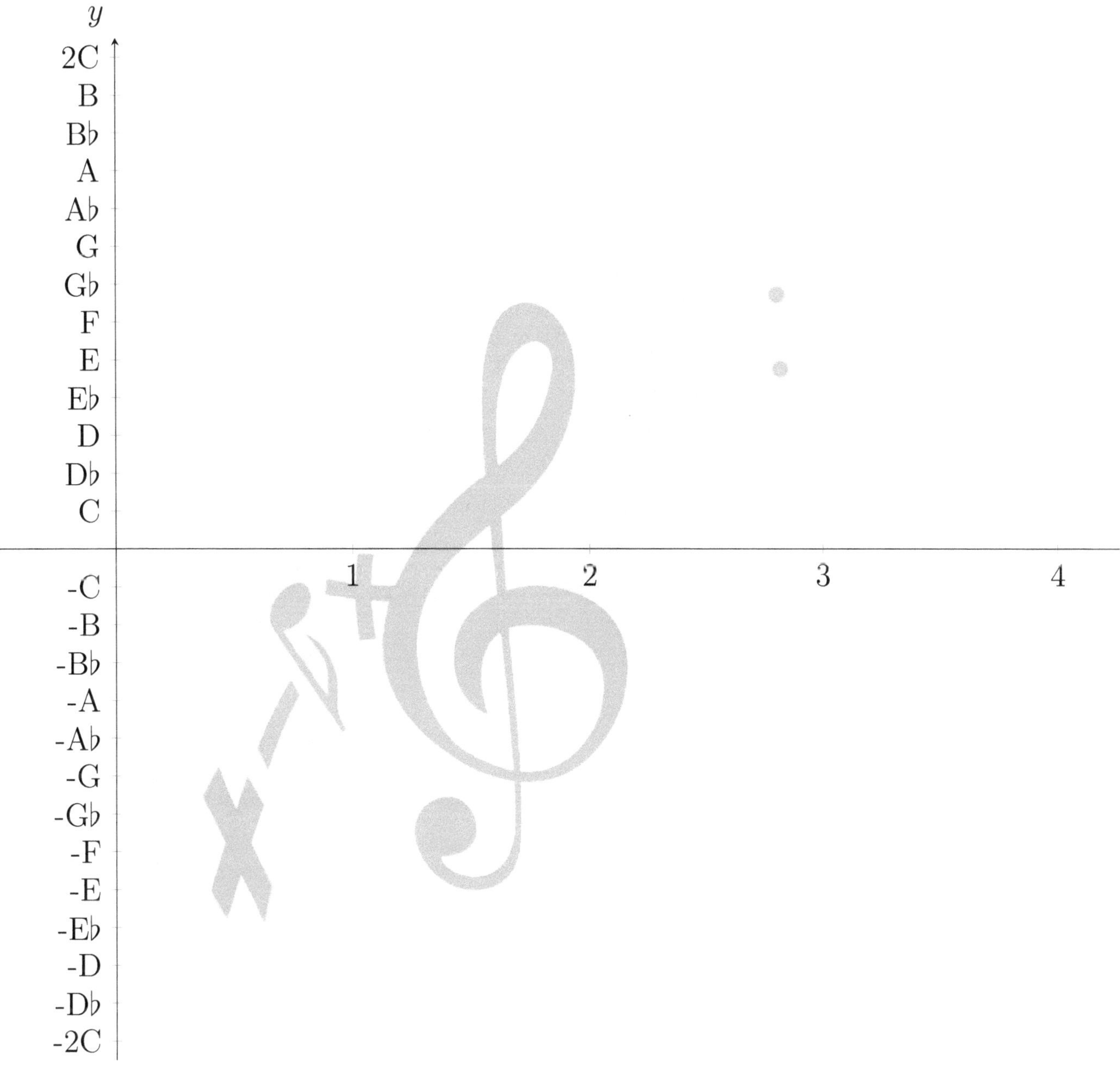

# Usher - Good Kisser

Please plot the following lines in order to play the featured song.

Line 1: (1, A), (2, G), (3, G), (4, F)

Line 2: (1, F), (2, E), (3, E♭), (4, C)

Line 3: (1, D), (2, D), (3, C), (4, B♭)

Line 4 : (1, C), (2, C), (3, B♭), (4, A)

Line 5 : (1, -G), (2, -G), (3, -D) (4, -D), (5, -C)

# Usher - Good Kisser

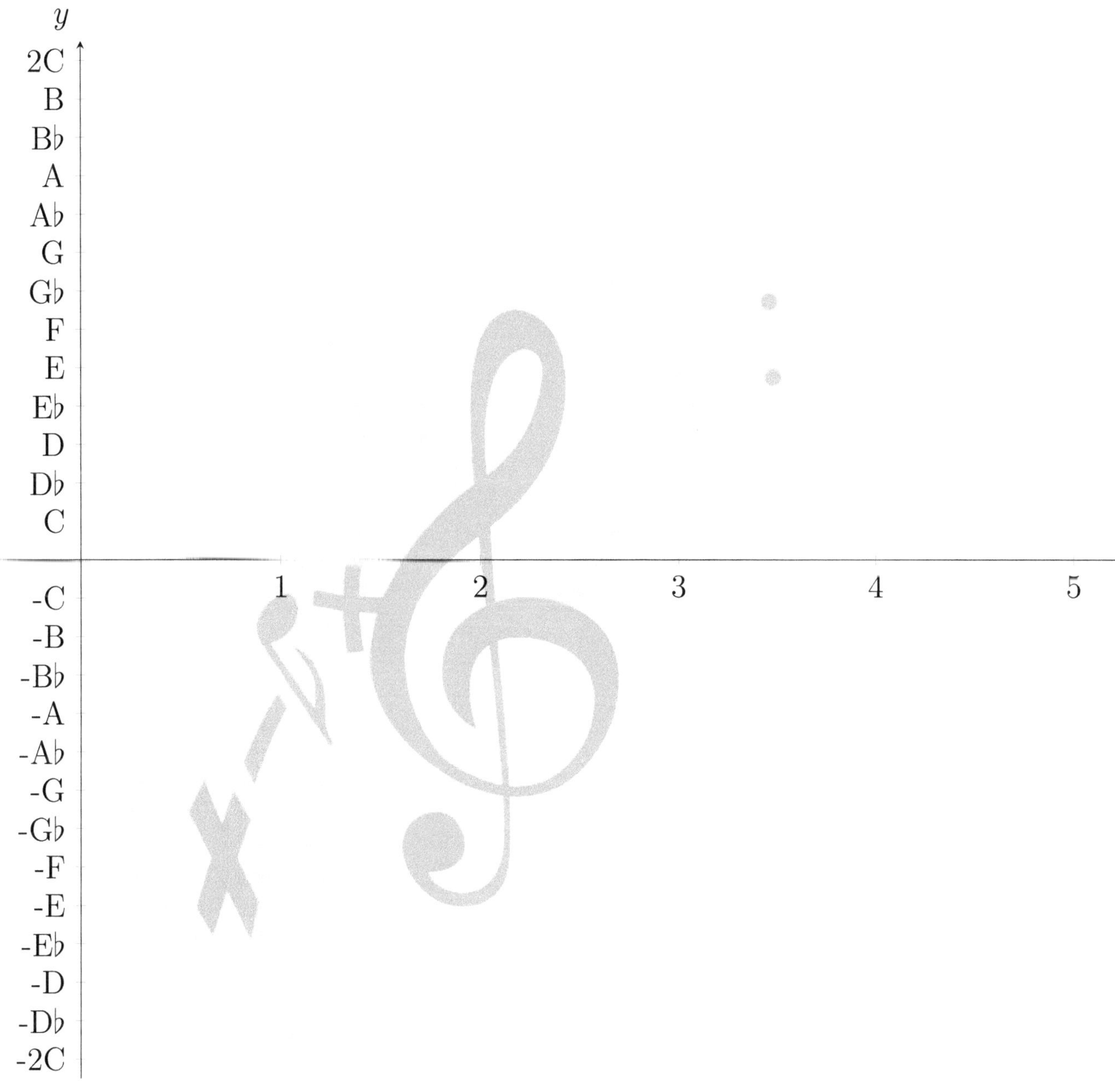

# Sam Smith - Stay With Me

Please plot the following lines in order to play the featured song.

Line 1: (1, A), (2, F), (3, E)

Line 2: (1, E), (2, C), (3, C)

Line 3: (1, C), (2, A), (3, G)

Line 4 : (1, -A), (2, -F), (3, -C)

# Sam Smith - Stay With Me

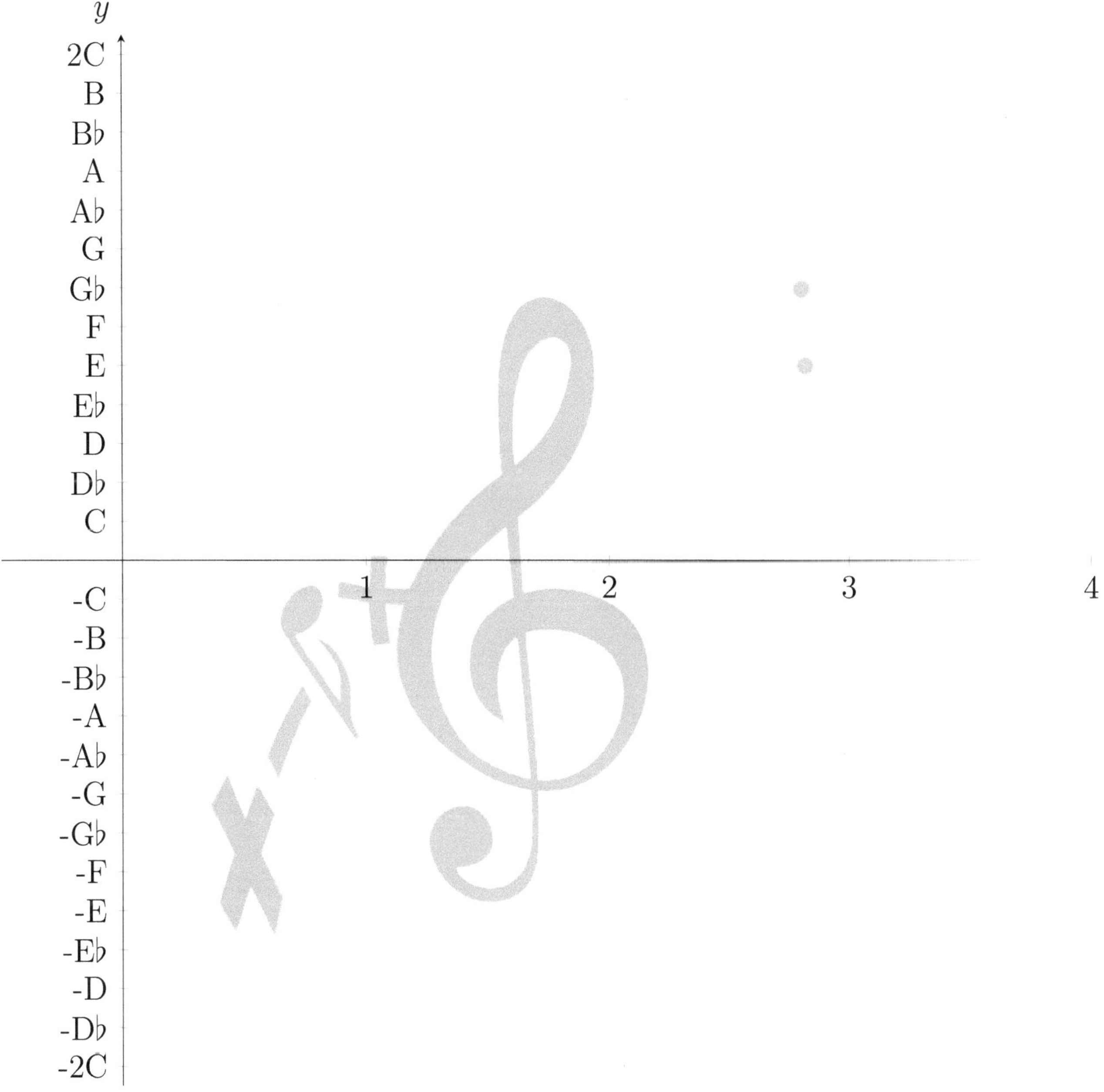

# Taylor Swift - Shake It Off

Please plot the following lines in order to play the featured song.

Line 1: (1, G), (2, G), (3, G)

Line 2: (1, E), (2, E), (3, D)

Line 3: (1, C), (2, C), (3, B)

Line 4 : (1, -A), (2, -C), (3, -G)

# Taylor Swift - Shake It Off

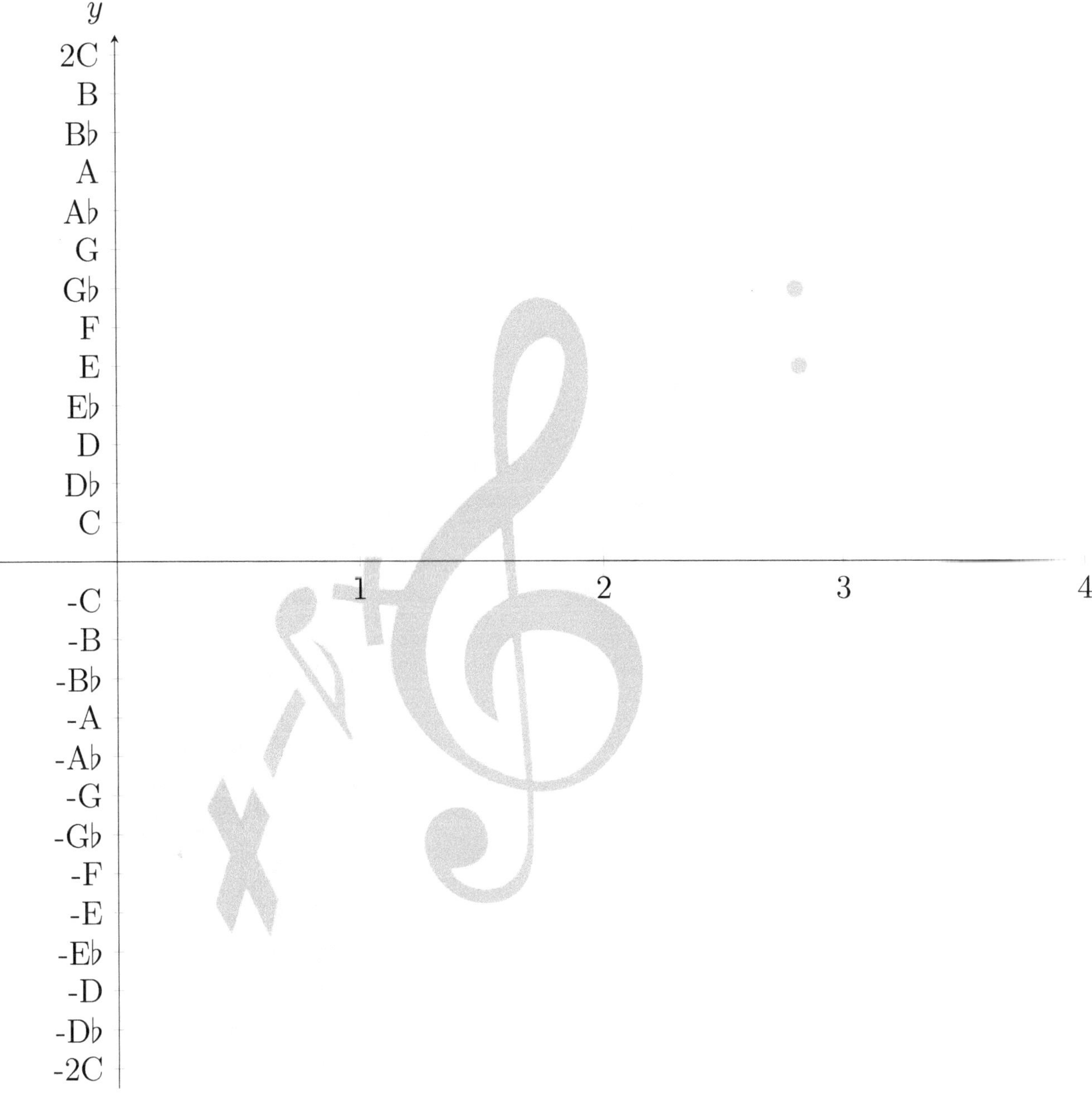

# 6 Slope-Intercept Form

During this section we will plot and draw lines that are in slope intercept form also seen as y = mx + b. We will use the same x and y axis as in the previous musical examples. The only difference that you will see in the slope intercept form is with "b" in the equation. the y intercepts or the "b" will always be a musical note.

For an example lets look at how to graph the line y = x + C♯

Please graph y = x + C♯

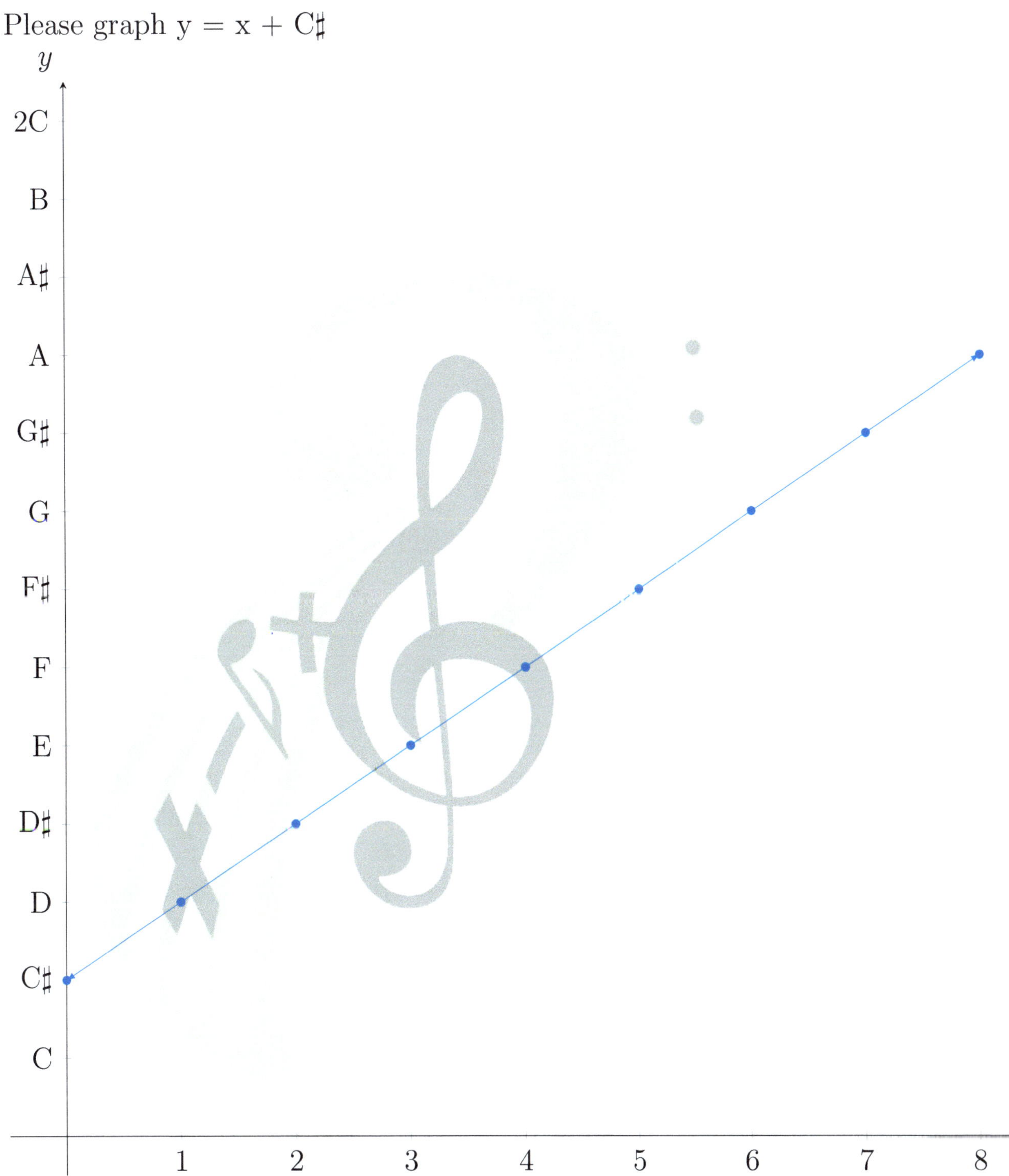

# Solution

Here is the solution to correctly graphing the line y = x + C♯. First lets note how the y intercept is C♯. This is the beginning point for graphing this line. The next step is to graph the remaining points by using the x as the slope. Be careful when plotting these points because the slope of x is $\frac{1}{1}$ but 1 is a point on the graph not a whole step. The eight musical notes that you would play on the piano from this line are D, D♯, E, F, F♯, G, G♯, A. Now of course this line is continuous in both directions but due to the graphs limitations we will only look at the eight notes plotted.

For the following examples solve the equations for y intercept form, graph and then play the results on the piano to hear what each line sounds like.

Graph the line and play the points to hear what it sounds like.

y = 2x + E

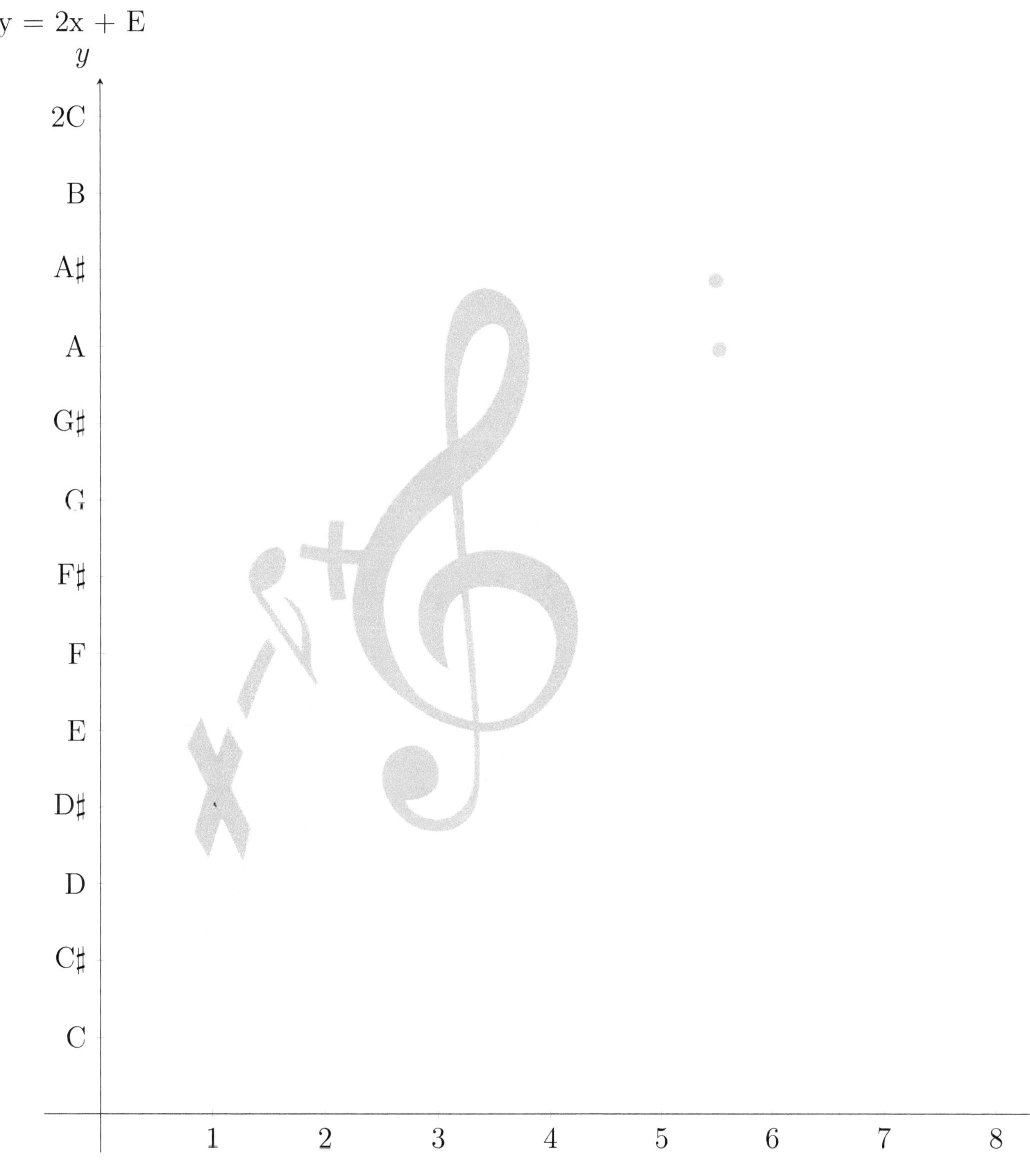

Graph the line and play the points to hear what it sounds like.

$y = -x + 2C$

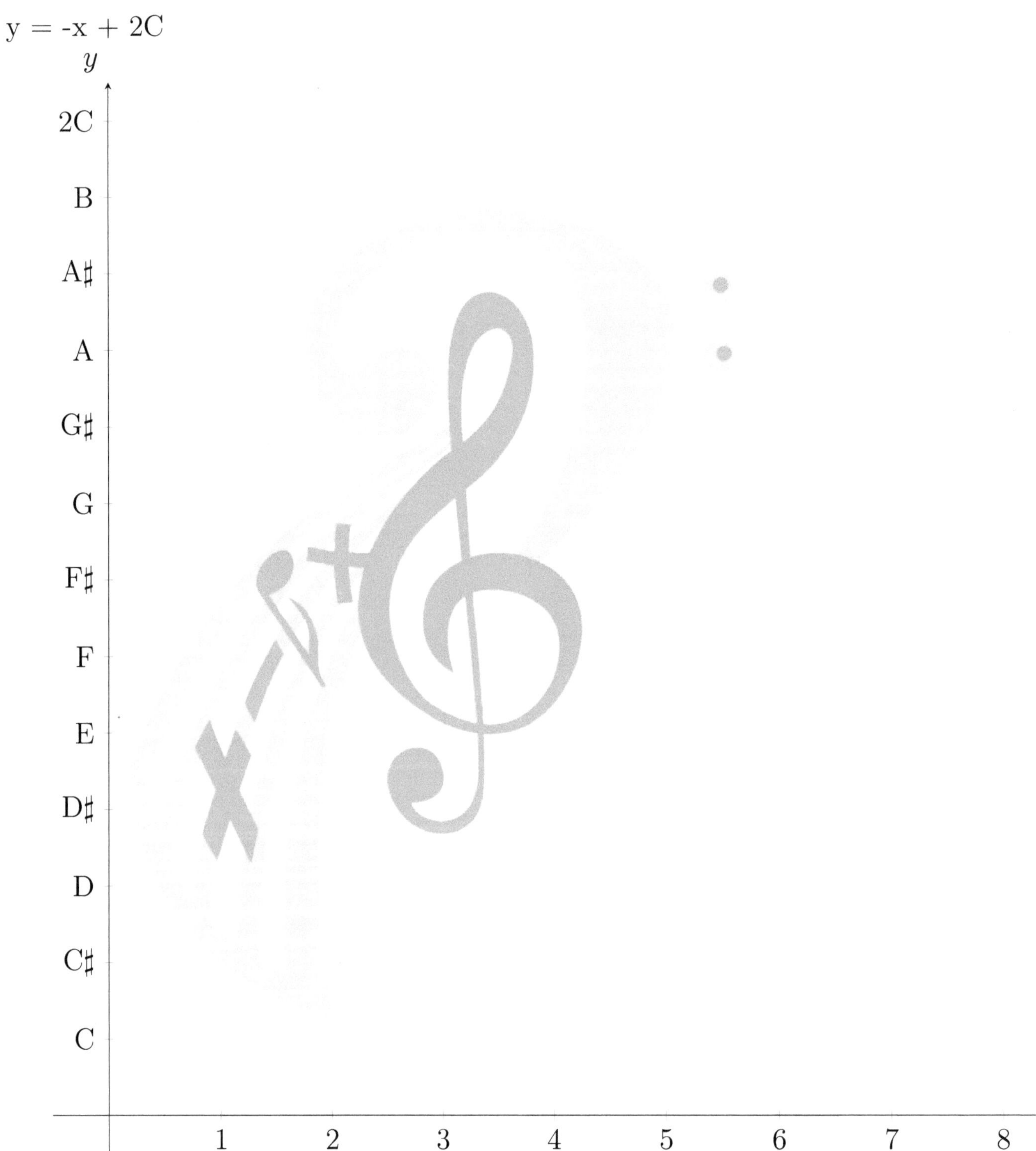

Graph the line and play the points to hear what it sounds like.

y = -3x + B

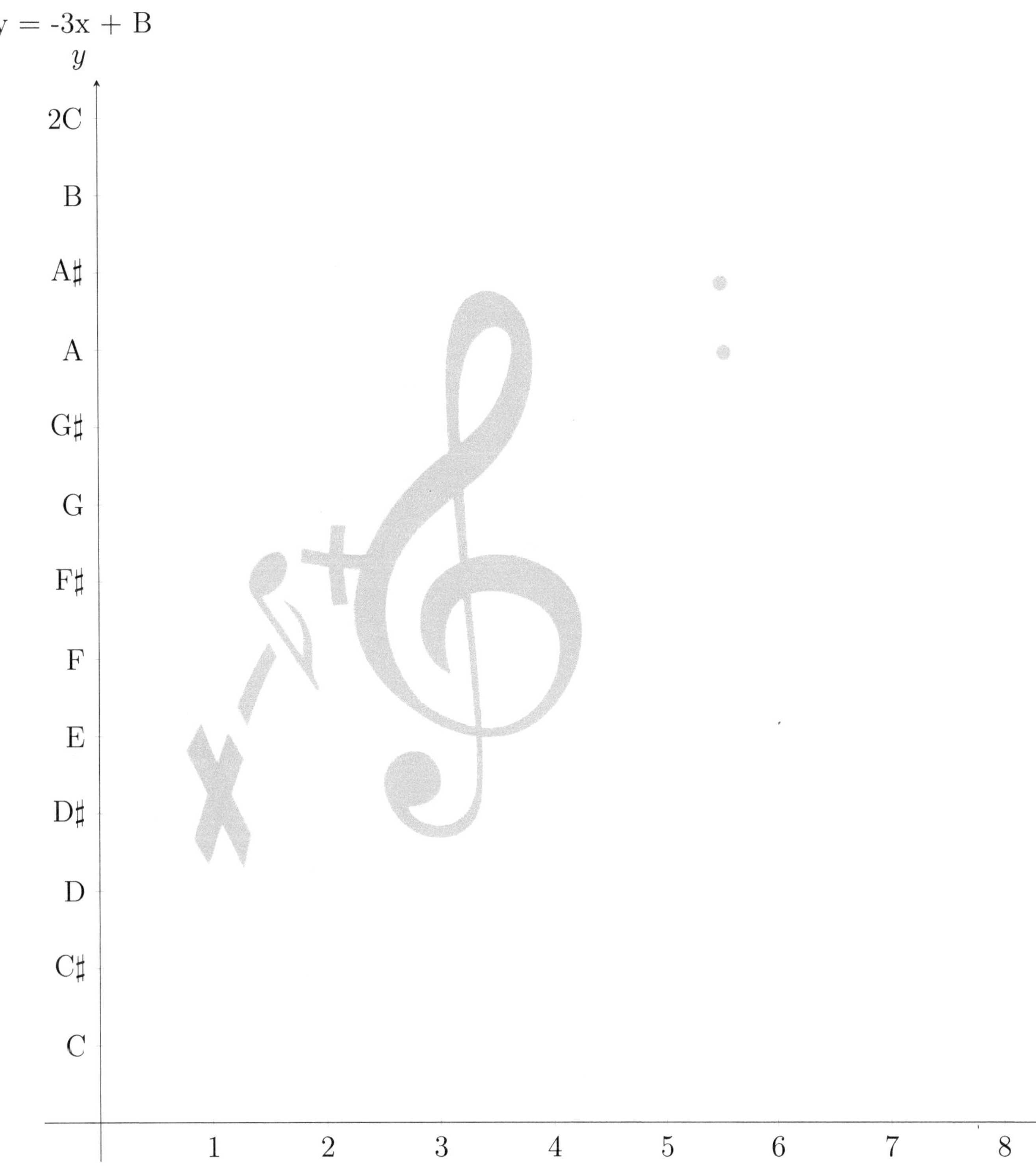

Graph the line and play the points to hear what it sounds like.

y = -$\frac{1}{2}$ + A♯

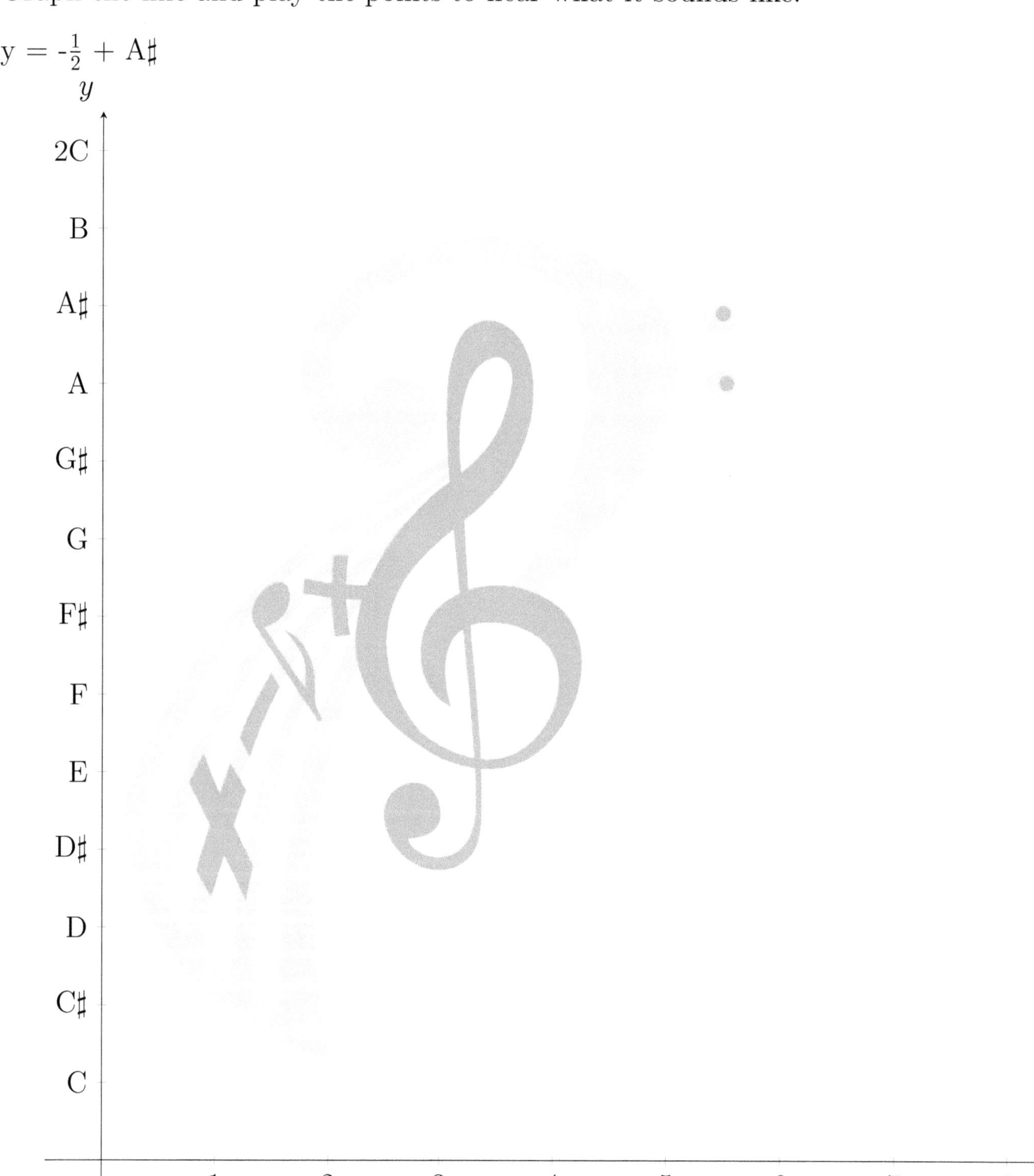

Graph the line and play the points to hear what it sounds like.

y = $\frac{1}{3}$

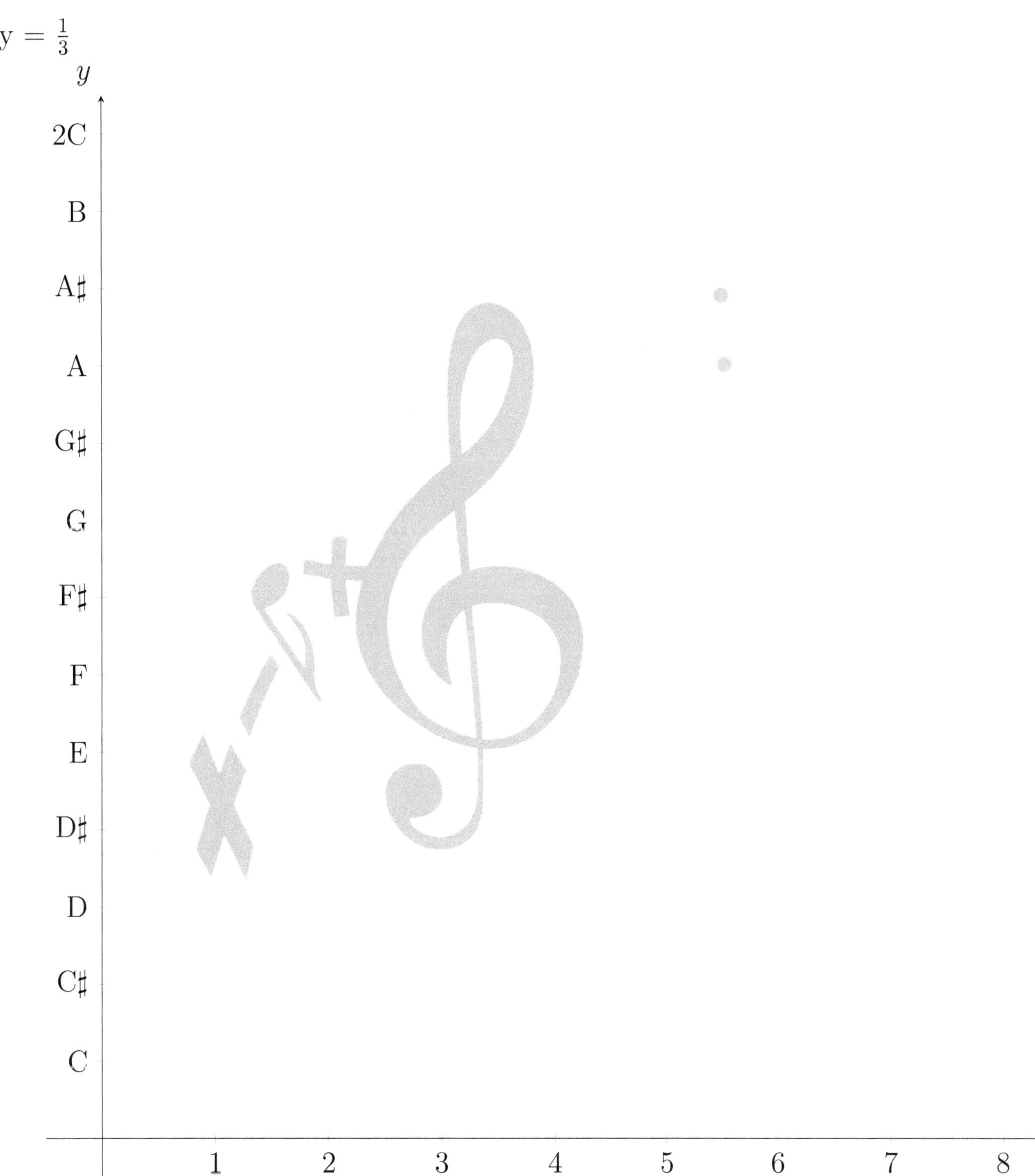

Graph the following lines and play the points to hear how each line sounds.

1. $y = 2x + C$
2. $y = \frac{1}{2}x + E$
3. $y = -x + B$

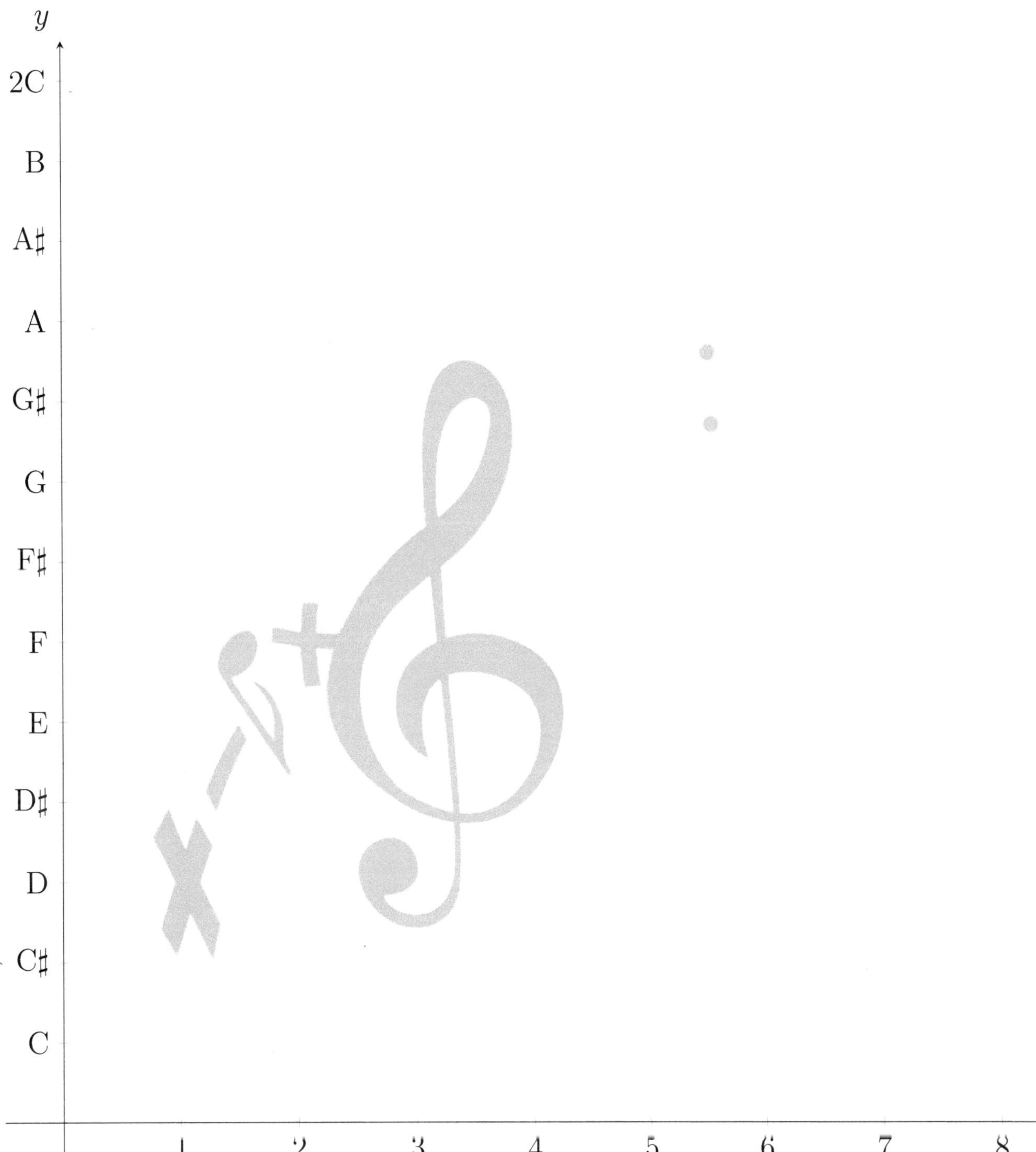
y
2C
B
A♯
A
G♯
G
F♯
F
E
D♯
D
C♯
C
1
2
3
4
5
6
7
8

Graph the following lines and play the points to hear how each line sounds.

1. y - 2x = -E♭

2. 2y + 4x = -2A

3. 3y - 3B♭ = -6x

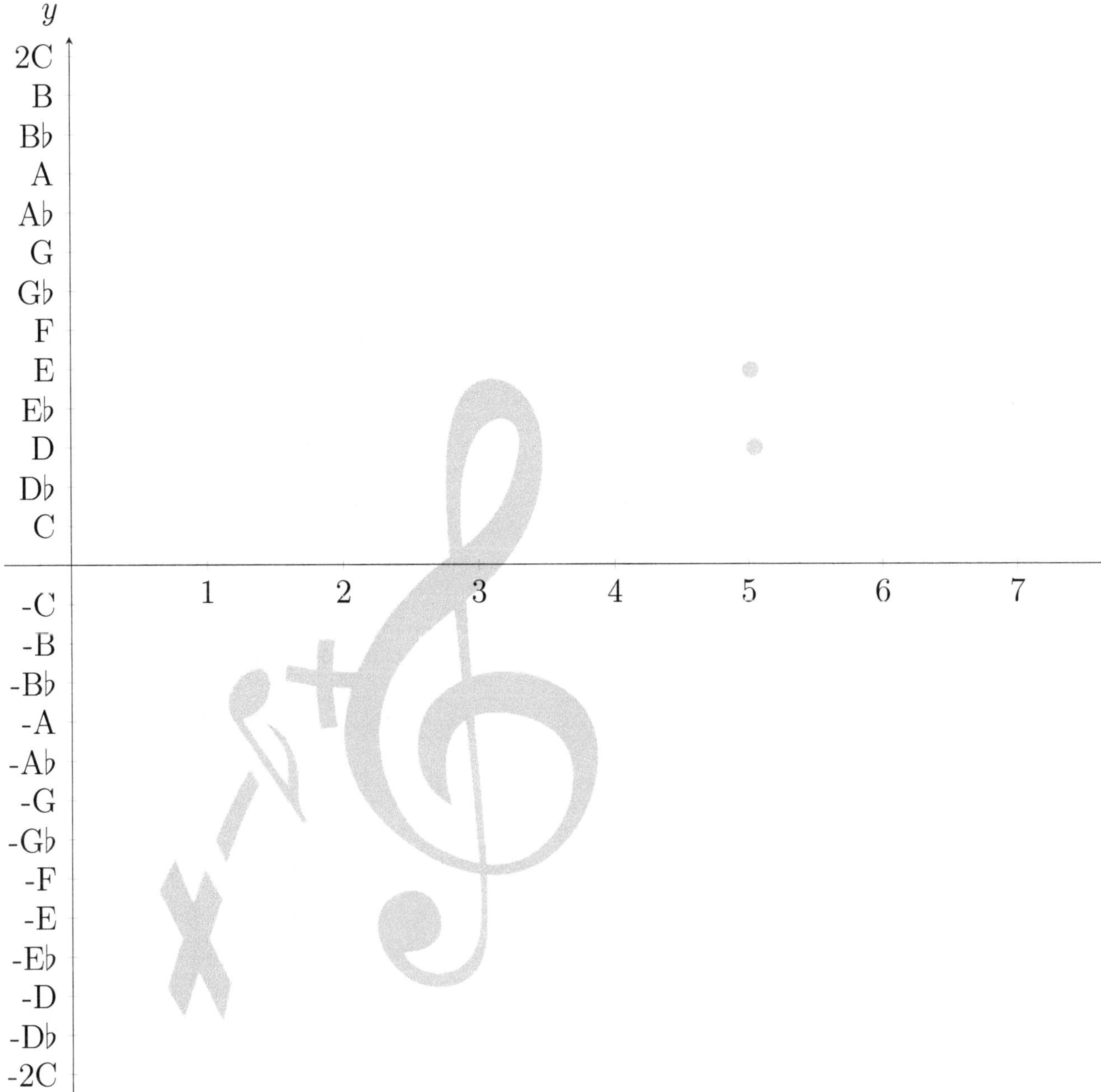
y
2C
B
B♭
A
A♭
G
G♭
F
E
E♭
D
D♭
C
-C
-B
-B♭
-A
-A♭
-G
-G♭
-F
-E
-E♭
-D
-D♭
-2C
1
2
3
4
5
6
7

Graph the following lines and play the points to hear how each line sounds.

1. $2y - 4C = -x$

2. $3y = 9x - 6C$

3. $y = 4x - G\flat$

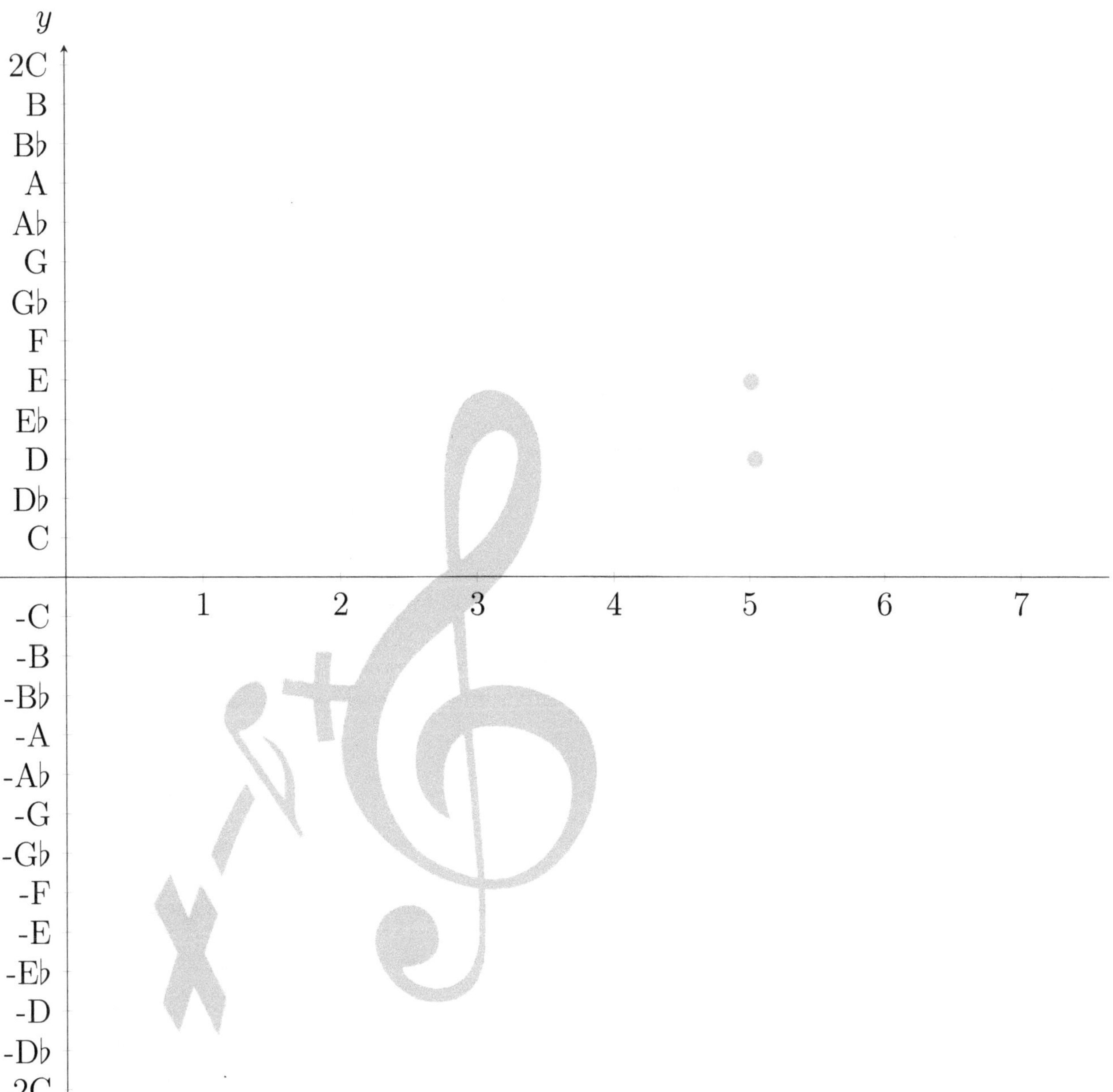
y
2C
B
B♭
A
A♭
G
G♭
F
E
E♭
D
D♭
C
-C
-B
-B♭
-A
-A♭
-G
-G♭
-F
-E
-E♭
-D
-D♭
-2C
1
2
3
4
5
6
7

Graph the following lines and play the points to hear how each line sounds.

1. $3y + 2x = 3A\flat$

2. $4y - 3x = -4E$

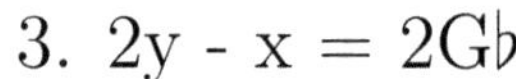

3. $2y - x = 2G\flat$

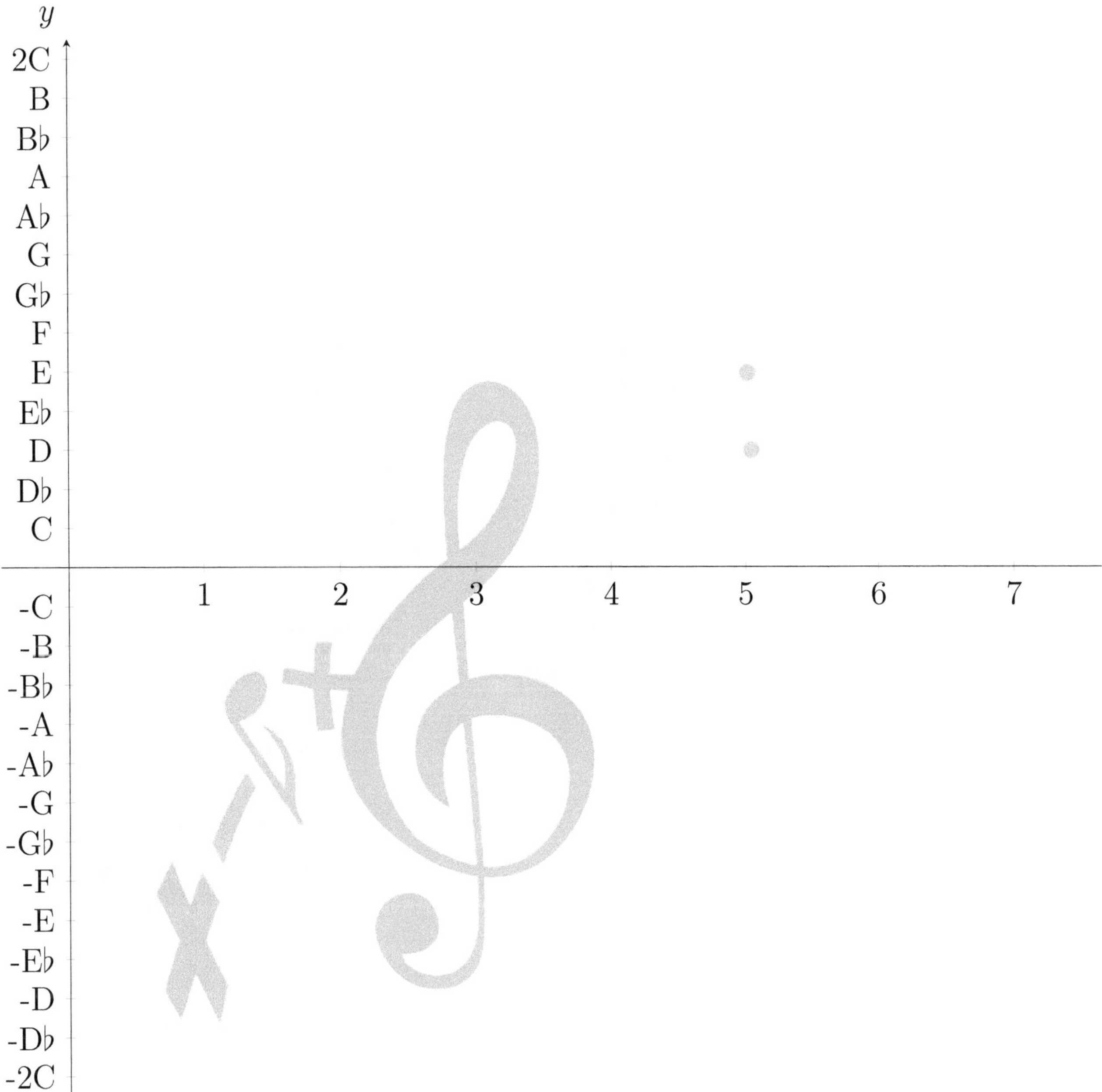
y
2C
B
B♭
A
A♭
G
G♭
F
E
E♭
D
D♭
C
-C
-B
-B♭
-A
-A♭
-G
-G♭
-F
-E
-E♭
-D
-D♭
-2C
1
2
3
4
5
6
7

Graph the following lines and play the points to hear how each line sounds.

1. $2y = 4x - 2B$

2. $3y + x = 3G$

3. $y = \frac{2}{3}x$

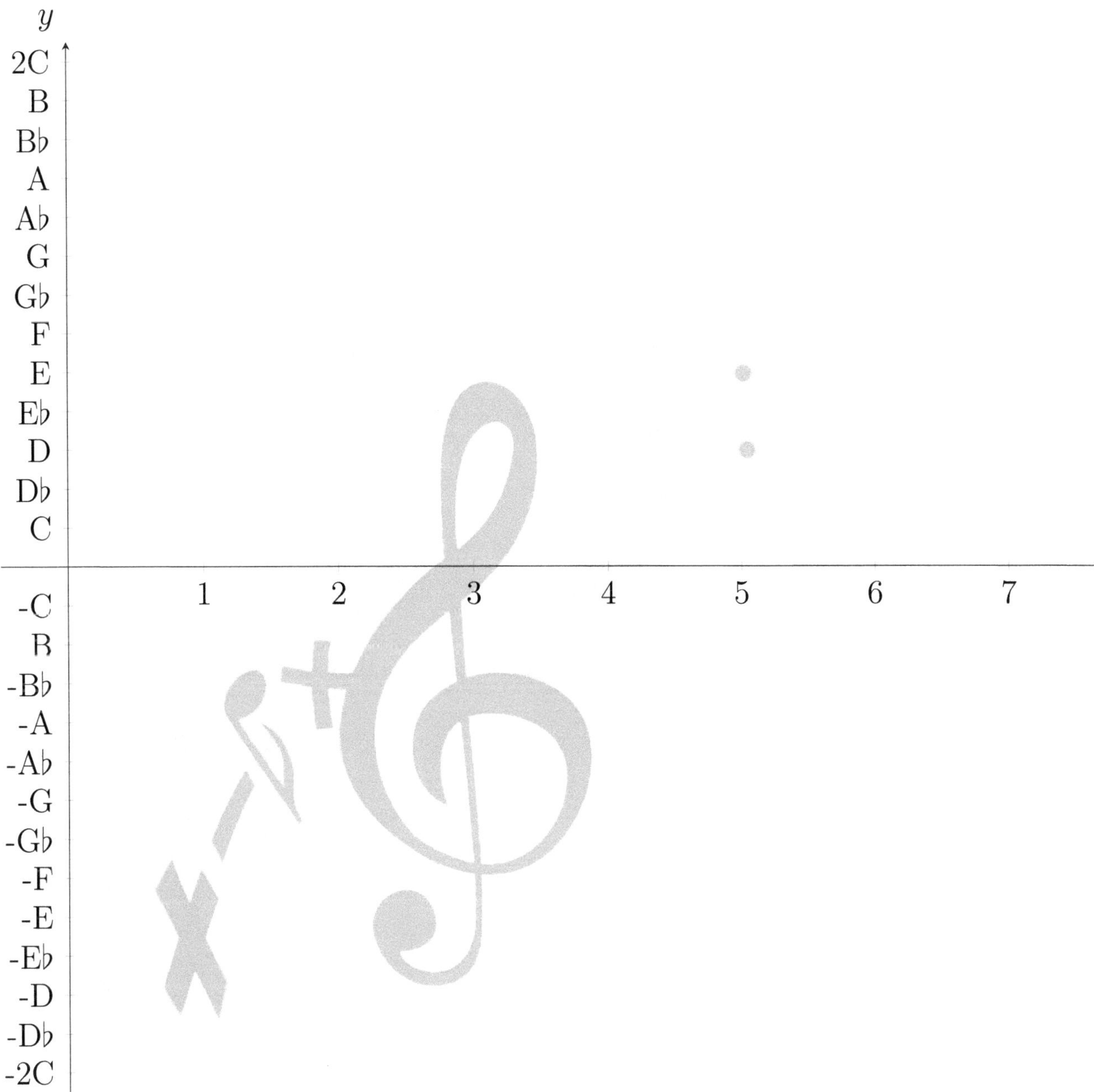
y
2C
B
B♭
A
A♭
G
G♭
F
E
E♭
D
D♭
C
1
2
3
4
5
6
7
-C
B
-B♭
-A
-A♭
-G
-G♭
-F
-E
-E♭
-D
-D♭
-2C

# 7 Solve By Substitution

In this section you will solve the following linear equations through substitution.

Solve the following equations through Substitution

3x - 2y = 7G♯

y = 5x

Solve the following equations through Substitution

$-3x + 5y = 5C$

$x + y = C$

Solve the following equations through substitution

-3x + 3y = 6F♯

y = -x

Solve the following equations through substitution

$2y - 4x = 6E\flat$

$2x = 4y$

Solve the following equations through substitution

x - 3y = 2C♯

3x = 3y

Solve the following equations through substitution

y + A♮ = x

2x - y = A♮

Solve the following equations through substitution

4x - 2y = 4F

2y = 4F

Solve the following equations through substitution

3y - x = -7C

y + 3x = C

# 8 Solve through Elimination

In this section you will solve for the following

Solve the following equations through elimination

4x - y = 5B

2y + x = -B

Solve the following equations through elimination

x + 3y = D♭

y - x = 3D♭

Solve the following equations through elimination

$3y + 3x = 9E$

$y - x = E$

Solve the following equations through elimination

$y - 2x = 2A\flat$

$2y + 2x = A\flat$

Solve the following equations through elimination

y + x = F♯

y - x = F♯

www.ingramcontent.com/pod-product-compliance
Lightning Source LLC
Chambersburg PA
CBHW050723180526
45159CB00003B/1110

*9781511772280*

## About The Authors

**Pankaj K. Godhaviya**

The author received his B.Sc., M.Sc. and Ph.D. degree in Chemistry, at the Saurashtra University at Rajkot. He has participated and presented scholarly research papers in national and international seminars. Four books are authored by him in Internationally reputed publishing house (1) Chalcones and Isoxazoles Scholars Press ISBN 9783639718799, (2) 1,2,4-Oxadiazoles - Scholars Press-ISBN-9783639661033, (3) Studies on 1,3-Thiazole derivatives (Biological Activity, Synthetic Aspects, Reaction Mechanism and Spectral Discussion) Lap Lambert Press-ISBN-9783659447051, (4) Fluorinating Reagents (Properties, Preparation, Application And Safety), CreateSpace Publishing ISBN-13: 9781511655248 and he also published four Research articles in internationally reputed journals. He has published 4 research papers in various reputed Journals. He is currently working as a Senior Research Associate in Research and Development (CRO) Division of Navin Fluorine International Limited, Surat, Gujarat (India). In past he had also worked with Jubilant Chemsys Limited, Noida (U.P.) (Medicinal chemistry division) and with Cadila Pharmaceuticals Limited, Ahmedabad (R & D-Peptide Chemistry Division).

**Contents** **Page No.**

**Abbreviations**

| | |
|---|---|
| $H_2F_2$ | Hydrogen fluoride |
| $Na_3AlF_3$ | Sodium aluminium fluoride |
| HF | Hydrofluoric acid |
| $CaF_2$ | Calsium fluoride |
| $KHF_2$ | Potassium hydrogen fluoride |
| $CF_2Cl_2$ | Dichlorodifluoromethane |
| $Et_3N$ | Triethyl Amine |
| HCl | Hydrochloric Acid |
| PFCs | Perfluorocarbons |
| PFOS | Perfluorooctanesulfonic acid |
| HCFCs | Hydrochlorofluorocarbons |
| PET | Positron emission tomography |
| PFAAs | Perfluoroalkylacid |
| NaOH | Sodium hydroxide |
| EtOH | Ethyl alcohol |
| $NH_3$ | Ammonia |
| KOH | Potassium hydroxide |
| $CDCl_3$ | Deuteriated Chloroform |
| DMSO | Dimethyl Sulfoxide |
| ppm | Part Per Million |
| DMF | Dimethyl Formamide |
| THF | Tetrahydrofuran |
| BuLi | Butyl Lithium |
| MHz | Mega Hertz |
| LD | Lethal Dose |
| KF | Potassium fluoride |
| DAST | Diethylaminosulfur trifluoride |
| MeCN | Acetonitrile |
| MeOH | Methyl alcohol |

# Chapter-1

# Element Fluorine

## 1. The Element Fluorine

9

**F**

Fluorine

18.9984032

### 1.1. Introduction:

#### 1.1.1. General propertics:

**Name:** Fluorine

**Symbol: Pronunciation:** FLOOR-een

**CAS Number:** 7782-41-4

#### 1.1.2. Fluorine in the periodic table:

**Atomic Number:** 9

**Atomic Weight:** 18.9984032

**Family:** Group 17 (VIIA) Halogen

**Period Number:** 2 **Group Number:** 17 **Group Name:** Halogen

**Element Classification:** Non-metal

**Electron Shell Configuration:**

$1s^2$

$2s^2$ $2p^5$

**Element category**: diatomic nonmetal

#### 1.1.3. Physical properties:

**Phase at Room Temperature:** Gas

**Appearance**

**gas:** very pale yellow
**liquid:** bright yellow
**solid:** alpha is opaque, beta is transparent

**Melting Point:** 53.53 K (-219.62°C or -363.32°F)

**Boiling Point:** 85.03 K (-188.12°C or -306.62°F)

**Liquid Density:** 0.001696 grams per cubic centimeter

**Triple point:** 53.48 K, 90 kPa

**Critical point:** 144.41 K, 5.1724 MPa

**Heat of vaporization:** 6.51 kJ·mol$^{-1}$

**Molar heat capacity:** $C_p$: 31 J·mol$^{-1}$·K$^{-1}$ (at 21.1 °C)
$C_v$: 23 J·mol$^{-1}$·K$^{-1}$ (at 21.1 °C)

**Vapor pressure:**

| P (Pa) | 1 | 10 | 100 | 1 k | 10 k | 100 k |
|---|---|---|---|---|---|---|
| at T (K) | 38 | 44 | 50 | 58 | 69 | 85 |

**1.1.4. Atomic properties:**

**Oxidation States:** -1

**Electronegativity:** 3.98 (Pauling scale)

**Ionization energies:** 1st: 1681 kJ·mol$^{-1}$
2nd: 3374 kJ·mol$^{-1}$
3rd: 6147 kJ·mol$^{-1}$

**Covalent radius:** 64 pm

**Van der Waals radius:** 135 pm

**Estimated Crustal Abundance:** $5.85\times10^2$ milligrams per kilogram

**Estimated Oceanic Abundance:** 1.3 milligrams per liter

### 1.1.5. Electron configuration:

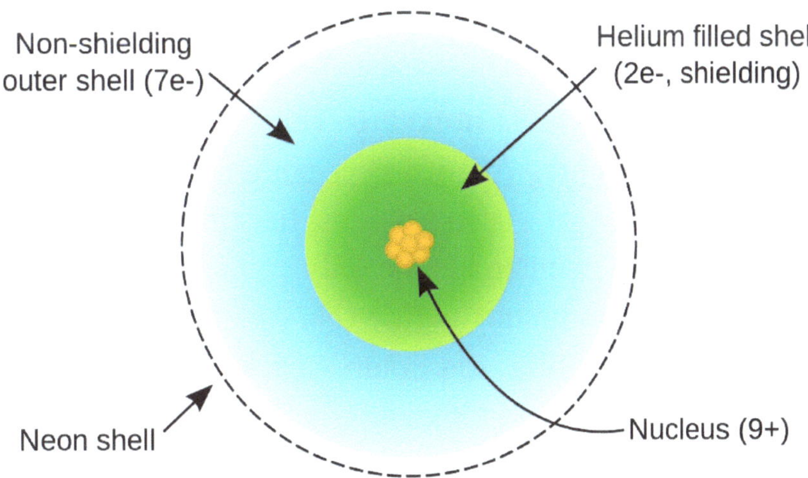


*Simplified structure of the fluorine atom*

Fluorine atoms have nine electrons, one fewer than neon, and electron configuration$1s^2 2s^2 2p^5$: two electrons in a filled inner shell and seven in an outer shell requiring one more to are filled. The outer electrons are ineffective at nuclear shielding, and experience a high effective nuclear charge of $9 - 2 = 7$; this affects the atom's physical properties.[1]

Fluorine's first ionization energy is third-highest among all elements, behind helium and neon,[2] which complicates the removal of electrons from neutral fluorine atoms. It also has a high electron affinity, second only to chlorine,[3] and tends to capture an electron to becomeisoelectronic with the noble gas neon;[1] it has the highest electronegativity of any element.[4] Fluorine atoms have a small covalent radius of around 60 picometers, similar to those of its period neighbours oxygen and neon.[5][6]

Fluorine is the lightest member of the halogen family, elements in Group 17 (VIIA) of the periodic table. The periodic table is a chart that shows how elements are related to one another. These include **chlorine, bromine, iodine,** and **astatine.** Fluorine is the most active chemical element, reacting with virtually every element. It even reacts with the noble gases at high temperatures and pressures. The noble gases, or Group 18 (VIIIA), also known as the inert gases, generally do not react with other elements.

Fluorine was discovered in 1886 by French chemist Henri Moissan (1852-1907). Moissan collected the gas by passing an electric current through one of its compounds, hydrogen fluoride ($H_2F_2$).

Consumers are most familiar with fluorine's use in two products. Fluorine gas is used to make fluorides, compounds that were made part of toothpastes in the 1950s. Fluorides are effective in preventing tooth decay and are added to urban water supplies as well.

Fluorine never occurs as a free element in nature. The most common fluorine minerals are fluorspar, fluorapatite, and cryolite. Apatite is a complex mineral containing primarily **calcium,** phosphorus, and oxygen, usually with fluorine. Cryolite is also known as Greenland spar. (The island of Greenland is the only commercial source of this mineral.) It consists primarily of sodium aluminum fluoride ($Na_3 ALF_6$). The major sources of these minerals are China, Mexico, Mongolia, and South Africa. The United States once produced small amounts of fluorspar, but its last remaining mine closed in 1995. The United States now imports the fluorine minerals it needs.

Fluorine is an abundant element in the Earth's crust, estimated at about 0.06 percent in the earth. That makes it about the 13th most common element in the crust. It is about as abundant as manganese or barium.

## 1.2. History

Fluorine is the most reactive of all elements and no chemical material is competent of release fluorine from any of its compounds. For this reason, fluorine does not come about free in nature and was very difficult for scientists to isolate.

### 1.2.1 Discovery and naming

Chemistry has always been a dangerous science. Early chemistry was a hazardous occupation. Men and women worked with chemicals about which they knew little. The discovery of new compounds and elements could easily have tragic consequences.

Flourine was particularly vicious. Chemists suffered terrible injuries and even died before the element was isolated. Fluorine gas is extremely damaging to the soft tissues of the respiratory tract.

In the early 1500s, German scholar Georgius Agricola (1494-1555) described a mineral he called fluorspar. The name fluorspar comes from the Latin word *fluere,* meaning "to flow." Agricola claimed that fluorspar added to molten metallic ores made them more liquid and easier to work with. Although Agricola did not realize it, fluorspar was a mineral of fluorine and contains calcium fluoride ($CaF_2$).

Fluorspar became the subject of intense study by early chemists. In 1670, German glass cutter Heinrich Schwanhard discovered that a mixture of fluorspar and acid formed a substance that could be used to etch glass. Etching is a process by which a pattern is drawn into glass. The chemical reaction leaves a frosted image. Etching is used to produce artistic shapes on glass as well as precise scientific measuring instruments.

The new etching material was identified in 1771 by Swedish chemist Carl Wilhelm Scheele (1742-86). Scheele described, in detail, the properties of this material, hydrofluoric acid (HF). His work set off an intense study of the acid and its composition.

One goal was to find ways to break hydrofluoric acid into elements. Chemists suspected that one element had never been seen before. Little did they know, however, what a dangerous new element it would be. During studies of hydrofluoric acid, many chemists were disabled when they inhaled hydrogen fluoride gas. One chemist, Belgian Paulin Louyet (1818-50), died from his exposure to the chemical. Finally, in 1888, the problem was solved. Moissan made a solution of hydrofluoric (HF) acid in potassium hydrogen fluoride ($KHF_2$). He then cooled the solution to -23°C (-9.4°F) and passed an electric current through it. A gas appeared at one end of the apparatus. He gave the name fluorine to the new element. The name comes from the mineral fluorspar.

Fluorine, poisonous in its elemental gaseous state, is also deadly in the common industrial chemical: hydrogen fluoride. Over a 10-year period, at least 10 people in the United States died from HF burns. Fluorine is also a component of "1080" poison, a mammal-killer banned in much of the world, but still used to kill Australian foxes and American coyotes. In dental products, when applied topically it chemically binds with the surface tooth enamel making it marginally more acid resistant. Although politically controversial, water fluoridation has shown consistent benefits, especially for poor children.

Because carbon-fluorine bonds are difficult to form, very few organisms make them. A few plants and bacteria in the tropics make fluorine-containing poisons. They do it to prevent being eaten. The same feature, the uncommon carbon-fluorine bond makes fluorination a powerful lever for new drug design. It allows tweaking organic molecules in a way nature never did and has led to several blockbuster commercial successes, such as Lipitor and Prozac.

Fluorine is added to city water supplies in the proportion of about one part per million to help prevent tooth decay. Sodium fluoride (NaF), stannous(II) fluoride ($SnF_2$) and sodium monofluorophosphate ($Na_2PO_3F$) are all fluorine compounds added to toothpaste, also to help prevent tooth decay. Hydrofluoric acid (HF) is used to etch glass, including most of the glass used in light bulbs. Uranium hexafluoride ($UF_6$) is used to separate isotopes of uranium. Crystals of calcium fluoride ($CaF_2$), also known as fluorite and fluorspar, are used to make lenses to focus infrared light. Fluorine joins with carbon to form a class of compounds known as fluorocarbons. Some of these compounds, such as dichlorodifluoromethane ($CF_2Cl_2$), were widely used in air conditioning and refrigeration systems and in aerosol spray cans, but have been phased out due to the damage they were causing to the earth's ozone layer.

In the environment, manmade fluorinated compounds have played roles in several noteworthy concerns. Chlorofluorocarbons were proven to damage the ozone layer and led to the wide-reaching Montreal Protocol treaty. (It was the chlorine in CFCs that was the bad actor, but fluorine was an important part of these molecules because it made them so stable and long-lived.) Another fluorine-related environmental issue is biopersistence. Long-lived molecules from waterproofing sprays, PFOA and PFOS, are found worldwide in wildlife and humans, including newborn children.

Fluorine biology also has its science fiction-y aspects. PFCs (perfluorocarbons) are capable of holding so much oxygen as to support human liquid breathing. Several works of fiction have touched on this, but in the real world, researchers have tried PFCs for burned lung care and as blood substitutes. Fluorine (in the form of its radioisotope F-18) is also at the heart of a modern imaging tenchique: positron emission tomography (PET). The technique produces 3-dimensional colored images of parts of the body that use a lot of sugar, particularly the brain or cancers.

Organofluorine chemistry began in the 1800s with the development of organic chemistry as a whole.[7] The first organofluorine compounds were prepared by metathesis reactions using antimony trifluoride as the $F^-$ source. The nonflammability and nontoxicity of the chlorofluorocarbons $CCl_3F$ and $CCl_2F_2$ attracted industrial attention in the 1920s. In the 1930s, scientists at duPont discovered polytetrafluoroethylene.[8] Subsequent major developments, especially in the US, benefited from expertise gained in the production of uranium hexafluoride.[9] Starting in the late 1940s, a series of electrophilic fluorinating methodologies were introduced, beginning with $CoF_3$. About this time, electrochemical fluorination ("electrofluorination") was announced, having been developed in the 1930s with the goal of generating highly stable perfluorinated materials compatible with uranium hexafluoride.[10] These new methodologies allowed the synthesis of C-F bonds without using elemental fluorine and without relying on metathetical methods. In 1957, the anticancer activity of 5-fluorouracil was described. This report provided one of the first examples of rational design of drugs.[11] This discovery sparked a surge of interest in fluorinated pharmaceuticals and agrichemicals. The discovery of the noble gas compounds, e.g. $XeF_4$, provided a host of new reagents starting in the early 1960s. In the 1970s, fluorodeoxyglucose was established as a useful reagent in $^{18}F$ positron emission tomography. In Nobel Prize-winning work, CFC's were shown to contribute to the depletion of atmospheric ozone. This discovery alerted the world to the negative consequences of organofluorine compounds and motivated the development of new routes to organofluorine compounds. In 2002, the first C-F bond-forming enzyme, fluorinase, was reported.[12]

### 1.3 Fluorine in different state:

#### 1.3.1 Fluorine gas

Elemental fluorine is highly toxic. Above a concentration of 25 ppm, it causes significant irritation while attacking the eyes, airways and lungs and affecting the liver and kidneys. At a concentration of 100 ppm, human eyes and noses are seriously damaged.[13]

#### 1.3.2 Hydrofluoric acid

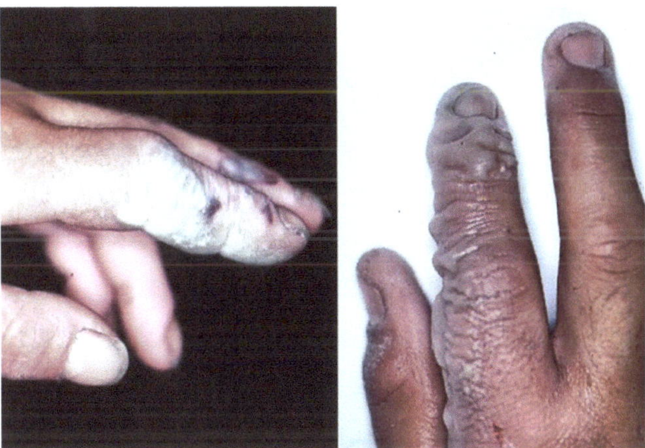

*Typical HF burns: the outward signs may not be evident for a day, at which point calcium treatments are less effective.*[14]

Hydrofluoric acid, the water solution of hydrogen fluoride, is a contact poison. Even though it is chemically only a weak acid, it is far more dangerous than the conventional strong mineral acids, such as nitric acid, sulfuric acid, or hydrochloric acid. Owing to its lesser chemical dissociation in water (remaining a neutral molecule), hydrogen fluoride penetrates tissue more quickly than typical acids. Poisoning can occur readily through the skin or eyes or when inhaled or swallowed. From 1984 to 1994, at least nine U.S. workers died from accidents with HF.[15]

Once in the blood, hydrogen fluoride reacts with calcium and magnesium, resulting in electrolyte imbalance (potentially hypocalcemia). The consequent effect on the heart (cardiac arrhythmia) may be fatal.[15] Formation of insoluble calcium fluoride also causes strong pain.[16] Burns with areas larger than 160 $cm^2$, about the size of a man's hand, can cause serious systemic toxicity.[17]

Symptoms of exposure to hydrofluoric acid may not be immediately evident, with 8-hour delay for 50% HF and up to 24 hours for lower concentrations. Hydrogen fluoride interferes with nerve function, meaning that burns may not initially be painful.

If the burn has been initially noticed, then HF should be washed off with a forceful stream of water for ten to fifteen minutes to prevent its further penetration into the body. Clothing used by the person burned may also present a danger.[18] Hydrofluoric acid exposure is often treated with calcium gluconate, a source of $Ca^{2+}$ that binds with the fluoride ions. Skin burns can be treated with a water wash and 2.5% calcium gluconate gel[19][20] or special rinsing solutions.[21] Because HF is absorbed, further medical treatment is necessary. Calcium gluconate may be injected or administered intravenously. Use of calcium chloride is contraindicated and may lead to severe complications. Sometimes surgical excision of tissue or amputation is required.[17][22]

### 1.3.3 Fluoride ion

Moroccan cow with fluorosis, from industrial contamination

Soluble fluorides are moderately toxic. For sodium fluoride, the lethal dose for adults is 5–10 g, which is equivalent to 32–64 mg of elemental fluoride per kilogram of body weight.[23] The dose that may lead to adverse health effects is about one fifth the lethal dose.[24] Chronic excess fluoride consumption can lead to skeletal fluorosis, a disease of the bones that affects millions in Asia and Africa.[24][25]

Historically, most cases of fluoride poisoning have been caused by accidental ingestion of insecticides containing inorganic fluoride.[26] Most calls to poison control centers for possible fluoride poisoning come from the ingestion of fluoride-containing toothpaste.[24] Malfunction of water fluoridation equipment has occurred several times, including an Alaskan incident that sickened nearly 300 people and killed one.[27]

## 1.4 Isotopes of fluorine

There is only one naturally occurring isotope of fluorine, fluroine-19. Isotopes are two or more forms of an element. Isotopes differ from each other according to their mass number. The number written to the right of the element's name is the mass number. The mass number represents the number of protons plus neutrons in the nucleus of an atom of the element. The number of protons determines the element, but the number of neutrons in the atom of any one element can vary. Each variation is an isotope.

Only one radioactive isotopes of fluorine, fluorine-18, has been prepared. A radioactive isotope is one that breaks apart and gives off some form of radiation. Radioactive isotopes are produced when very small particles are fired at atoms. These particles stick in the atoms and make them radioactive.

Fluroine-18 is sometimes used for medical studies. It is injected into the body where it travels primarily to bones. Its presence in bones can be detected by the radiation it gives off. The radiation pattern discloses how normal bones are. Fluorine-18 is sometimes used in a similar way to study brain function.

Although **fluorine (F)** has 18 known **isotopes** from $^{14}F$ to $^{31}F$ and one isomer ($^{18m}F$), only one of these isotopes is stable, that is, fluorine-19; as such, it is a monoisotopic element. The longest-lived radioisotope is $^{18}F$ with a half-life of 109.771 minutes. All other isotopes have half-lives under a minute, the majority under second, making fluorine a mononuclidic element as well. The least stable isotope is $^{15}F$, whose half-life is 4.1 x $10^{-22}$ seconds, corresponding to a spectral linewidth of about 1 MeV. Only $^{14}F$ has an unknown half-life.

Standard atomic mass: 18.9984032(5) u

**Fluorine-18**

The nuclide $^{18}F$ is the radionuclide of fluorine with the longest half-life, 109.771 minutes, allowing it to serve commercially as an important source of positrons. Its major use is for the production of the radiopharmaceutical fludeoxyglucose for positron emission tomography scanning in medicine.

Like all positron-emissing radioisotopes, $^{18}F$ also has a probability to decay by electron capture. In this case, $^{18}F$ decays into $^{18}O$, 96.86 (19)% of the time by beta plus (positron) emission and 3.14 (19)% by electron capture.

It is the lightest unstable nuclide with equal odd numbers of protons and neutrons, 9 of each. (See also the "magic numbers" discussion of nuclide stability.)

**Fluorine-19**

Fluorine-19, the only stable isotope of fluorine. Its abundance is 100%; no other isotopes of fluorine exist in significant quantities. Its binding energy is 147801 keV. Fluorine-19 is NMR-active, so it is used in fluorine-19 NMR spectroscopy.

**Fluorine-20**

Fluorine-20, is one of the more unstable isotopes of fluorine. It has a half-life of 11.07 seconds and undergoes beta decay, transforming into its daughter nuclide $^{20}Ne$. Its specific radioactivity is 1.885 x $10^{9}$ TBq/g and has a lifetime of 15.87 seconds.

**Fluorine-21**

Fluorine-21, as with fluorine-20, is also one of the more unstable isotopes of this element. It has a half-life of 4.158 seconds. It undergoes beta decay as well which leaves behind a daughter nuclei of $^{21}Ne$. Its specific radioactivity is $4.78 \times 10^9$ TBq/g.

## 1.5 Source of fluorine:

### 1.5.1 Natural biochemistry

*South Africa's gifblaar is one of the few organisms that make fluorine compounds.*

Biologically synthesized organofluorines have been found in microorganisms and plants,[28] but not in animals.[29] The most common example is fluoroacetate, with an active poison molecule identical to commercial "1080". It is used as a defence against herbivores by at least 40 green plants in Australia, Brazil, and Africa;[30] other biologically synthesized organofluorines include ω-fluoro fatty acids, fluoroacetone, and 2-fluorocitrate.[29] In bacteria, the enzyme adenosyl-fluoride synthase, which makes the carbon–fluorine bond, was isolated. The discovery was touted as possibly leading to biological routes for organofluorine synthesis.[31]

Fluoride is not considered an essential mineral element for mammals and humans.[32] Small amounts of fluoride may be beneficial for bone strength, but this is an issue only in the formulation of artificial diets.[33]

### 1.5.2 Chemical synthesis:

While preparing for a 1986 conference to celebrate the centennial of Moissan's achievement, Karl O. Christe reasoned that chemical fluorine generation should be feasible since some metal fluoride anions have no stable neutral counterparts; their acidification potentially triggers oxidation instead. He devised a method which evolves fluorine at high yield and atmospheric pressure:

$$2\ KMnO_4 + 2\ KF + 10\ HF + 3\ H_2O_2 \rightarrow 2\ K_2MnF_6 + 8\ H_2O + 3\ O_2\uparrow$$

$$2\ K_2MnF_6 + 4\ SbF_5 \rightarrow 4\ KSbF_6 + 2\ MnF_3 + F_2\uparrow$$

Christe later commented that the reactants "had been known for more than 100 years and even Moissan could have come up with this scheme. As late as 2008, some references still asserted that fluorine was too reactive for any chemical isolation.

Moissan's method is used to produce industrial quantities of fluorine, via the electrolysis of a potassium fluoride/hydrogen fluoride mixture: hydrogen and fluoride ions are reduced and oxidized at a steel container cathode and a carbon block anode, under 8–12 volts, to generate hydrogen and fluorine gas respectively. Temperatures are elevated, KF•2HF melting at 70 °C (158 °F) and being electrolyzed at 70–130 °C (158–266 °F). KF, which acts as catalyst, is essential since pure HF cannot be electrolyzed. Fluorine can be stored in steel cylinders that have passivated interiors, at temperatures below 200 °C (392 °F); otherwise nickel can be used. Regulator valves and pipe work are made of nickel, the latter possibly using Monel instead. Frequent passivation, along with the strict exclusion of water and greases, must be undertaken. In the laboratory, glassware may carry fluorine gas under low pressure and anhydrous conditions; some sources instead recommend nickel-Monel-PTFE systems.

Elemental fluorine's large-scale synthesis began during World War II. Germany used high-temperature electrolysis to make tons of the planned incendiary chlorine trifluoride and the Manhattan Project used huge quantities to produce uranium hexafluoride for uranium enrichment. Since $UF_6$ is as corrosive as fluorine, gaseous diffusion plants required special materials: nickel for membranes, fluoropolymers for seals, and liquid fluorocarbons as coolants and lubricants. This burgeoning nuclear industry later drove post-war fluorochemical development.

## 1.6 Fluorine chemical Properties and reactivity:

### 1.6.1 Chemical properties

Fluorine is the most reactive element. It combines easily with every other element except **helium, neon,** and **argon.** It reacts with most compounds, often violently. For example, when mixed with water, it reacts explosively. For these reasons, it must be handled with extreme care in the laboratory.

### 1.6.2 Reactivity:

The bond energy of difluorine is much lower than that of either $Cl_2$ or $Br_2$ and similar to the easily cleaved peroxide bond; this, along with high electronegativity, accounts for fluorine's easy dissociation, high reactivity and strong bonds to non-fluorine atoms.[34][35] Conversely, bonds to other atoms are very strong because of fluorine's high electronegativity. Unreactive substances like powdered steel, glass fragments, and asbestos fibers react quickly with cold fluorine gas; wood and water spontaneously combust under a fluorine jet.[36][37]

Reactions of elemental fluorine with metals require varying conditions. Alkali metals cause explosions and alkaline earth metals display vigorous activity in bulk; to prevent passivation from the formation of metal fluoride layers most other metals such as aluminium and iron must be powdered,[34] and noble metals require pure fluorine gas at 300–450 °C (575–850 °F).[38] Some solid nonmetals (sulfur, phosphorus) react vigorously in liquid air temperature fluorine.[39] Hydrogen sulfide[39] and sulfur dioxide[40] combine readily with fluorine, the latter sometimes explosively; sulfuric acidexhibits much less activity, requiring elevated temperatures.[41]

Hydrogen, like the alkali metals, reacts explosively with fluorine.[42] Carbon, as lamp black, reacts at room temperature to yield fluoromethane. Graphite combines with fluorine above 400 °C (750 °F) to produce non-stoichiometric carbon monofluoride; higher temperatures generate gaseous fluorocarbons, sometimes with explosions.[43] Carbon dioxide and carbon monoxide react at or just above room temperature,[44] whereas paraffins and other organic chemicals generate strong reactions:[45] even fully substituted haloalkanes such as carbon tetrachloride, normally incombustible, may explode.[46] Although nitrogen trifluoride is stable, nitrogen requires an electric discharge at elevated temperatures for reaction due to its very strong triple bond;[47] ammonia may react explosively.[48][49] Oxygen does not combine with fluorine under ambient conditions, but can be made to using electric discharge at low temperatures and pressures; the products tend to disintegrate into their constituent elements when heated.[50][51][52] Heavier halogens[53] and radon[54] react readily with fluorine; the lighter noble gases xenon and krypton require special conditions.[55]

## 1.7 Fluoride Introduction

Fluoride is a naturally occurring chemical ion of the element *fluorine*. Fluoride has one extra electron than fluorine giving it a negative charge. It can be found in water, food, soil and different minerals like *fluorite* and *fluorapatite*. It is also artificially created in laboratories to be used for a variety of reasons.

When people think of fluoride, they most commonly think of dental hygiene products such as various mouthwashes or toothpastes.

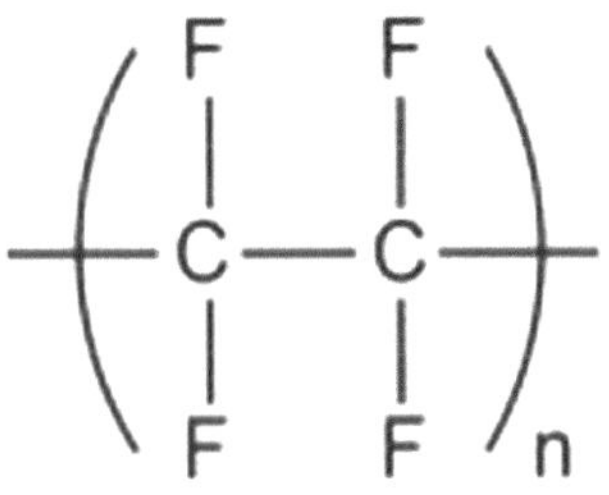

*Chemical Structure of Fluoride*

Fluoride works in two ways. The combination of sugar and mouth bacteria creates an acid that erodes enamel and can damage teeth. Fluoride can defend teeth from the *demineralization* that is caused by this acid. If teeth have already been damaged by this acid, fluoride collects in the demineralized areas and starts to strengthen enamel. This is called *remineralization.* Fluoride is very efficient and useful for preventing cavities; however its effects may be in vain if cavities have already formed.

The straight answer is no. When used in moderation, it can be effective in preventing tooth decay. If used in high levels over long periods of time, it can be extremely harmful to overall health and wellbeing. One example is dental fluorosis, which is a discoloration of tooth enamel. It can also weaken bones and lead to skeletal fluorosis which is most commonly associated with joint stiffness and joint pain very similar to that of arthritis.

In extreme cases (like if someone were to ingest an entire tube of toothpaste), fluoride can be extremely toxic. An overdose can cause symptoms including nausea, diarrhea, stomach pain/discomfort, salivation, vomiting blood, watering of the eyes, body weakness, shallow breath, faintness, tiredness, convulsions, and even death.

**What should you do if you or someone you know overdoses on fluoride?**

1. Call Poison Control: The National Poison Control Center (1-800-222-1222) can be called from anywhere in the United States.
2. Determine the patient's age, weight, and condition (for example, is the person conscious?) Name of the product (ingredients and strengths, if known), time it was swallowed and the amount swallowed.
3. Seek out emergency assistance immediately. A trip to the ER could save your life.

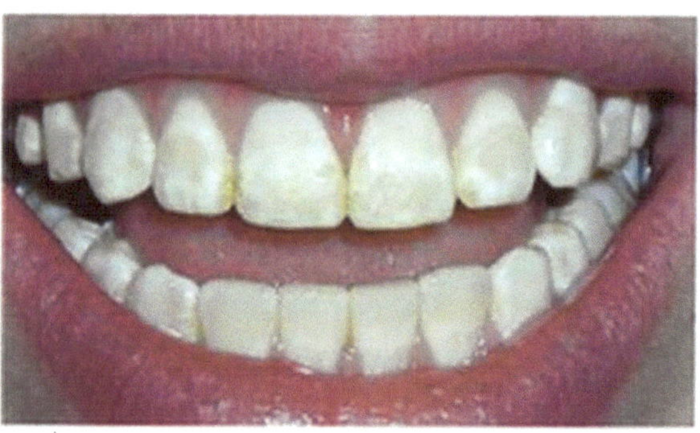

*A mild case of Dental Fluorosis*

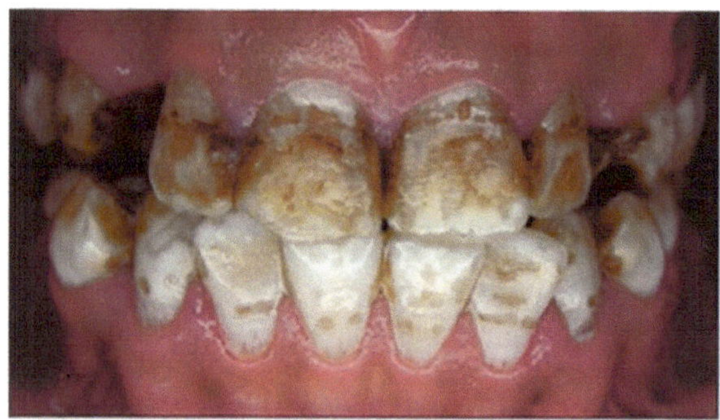

*An extreme case of Dental Fluorosis*

## 1.8. Environment concern:

### 1.8.1 Atmosphere

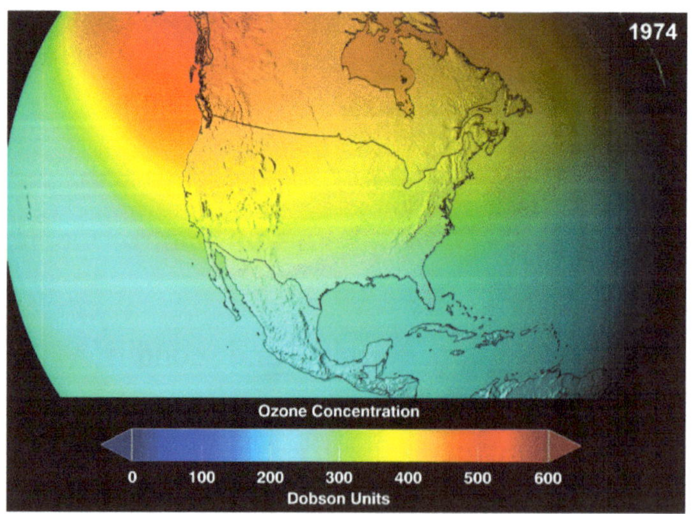


*NASA projection of stratospheric ozone over North America without the Montreal Protoco*

The Montreal Protocol, signed in 1987, set strict regulations on chlorofluorocarbons (CFCs) and bromofluorocarbons due to their ozone damaging potential (ODP). The high stability which suited them to their original applications also meant that they were not decomposing until they reached higher altitudes, where liberated chlorine and bromine atoms attacked ozone molecules.[56] Even with the ban, and early indications of its efficiency, predictions warned that several generations would pass before full recovery.[57][58] With one-tenth the ODP of CFCs, hydrochlorofluorocarbons (HCFCs) are the current replacements,[59] and are themselves scheduled for substitution by 2030–2040 by hydrofluorocarbons (HFCs) with no chlorine and zero ODP.[60] In 2007 this date was brought forward to 2020 for developed countries;[61] the Environmental Protection Agency had already prohibited one HCFC's production and capped those of two others in 2003.[60] Fluorocarbon gases are generally greenhouse gases with global-warming potentials (GWPs) of about 100 to 10,000; sulfur hexafluoride has a value of around 20,000.[62] An outlier is HFO-1234yf which has attracted global demand due to its GWP of 4 compared to 1,430 for the current refrigerant standard HFC-134a.[63]

### 1.8.2 Biopersistence

Because of the strength of the carbon–fluorine bond, organofluorines endure in the environment. Perfluoroalkyl acids (PFAAs) have attracted particular attention as persistent global contaminants. These compounds can enter the environment from their direct uses in waterproofing treatments and firefighting foams or indirectly from leaks from fluoropolymer production plants (where they are intermediates). Because of the acid group, PFAAs are water soluble in low concentrations.[64] While there are other PFAAs, the lion's share of environmental research has been done on the two most well-known: perfluorooctanesulfonic acid (PFOS) and perfluorooctanoic acid (PFOA). The U.S. Environmental Protection Agency classifies these materials as "emerging contaminants" based on the growing but still incomplete understanding of their environmental impact.[65][66][67]

Trace quantities of PFAAs have been detected worldwide, from polar bears in the Arctic to the global human population. Both PFOS and PFOA have been detected in breast milk and the blood of newborns. A 2013 review showed widely varying amounts of PFOS and PFOA in different soils and groundwater, with no clear pattern of one chemical dominating. PFAA concentration was generally higher in areas with more human populations or industrial activity, and areas with more PFOS generally also had more PFOA.[68] Human populations also showed different concentrations of the two chemicals; for example one study showed more PFOS than PFOA in Germans, while another study showed the reverse for Americans. PFAAs may be starting to decrease in the biosphere: one study indicated that PFOS levels in wildlife in Minnesota were going down, presumably because of ceased production of the chemical by 3M.[65][66]

*The PFOS molecule*

In the body, PFAAs bind to proteins such as serum albumin. Their tissue distribution in humans is unknown, but studies in rats suggest it is present mostly in the liver, kidney, and blood. They are not metabolized by the body but are excreted by the kidneys. Dwell time in the body varies greatly by species. Rodents have half-lives of days, while in humans they remain for years. Many animals show sex differences in the ability to rid the body of PFAAs, but without a clear pattern. It varies by the animal if half lives are longer for males or females.[65][66][69]

The potential health impact of PFAAs is unclear. Unlike chlorinated hydrocarbons, PFAAs are not lipophilic (stored in fat), nor genotoxic (damaging genes). Both PFOA and PFOS in high doses cause cancer and the death of newborns in rodents. However, studies on humans have not been able to prove an impact at current exposures. Bottlenose dolphins have some of the highest PFOS concentrations of any wildlife studied; one study suggests an impact on their immune systems.[65][66][69]

The biochemical causes of toxicity are also unclear and may differ by molecule, health effect, and even animal. Significant research has been done looking at PPAR-alpha (a protein that interacts with PFAAs and is commonly implicated in contaminant-caused rodent cancers.[65][66][69]

Less fluorinated chemicals (not perfluorinated compounds) are also detectable in the environment. Because biological systems do not metabolize fluorinated molecules easily, fluorinated pharmaceuticals (often antibiotics and antidepressants) are among the major fluorinated organics found in treated city sewage and wastewater.[70] Fluorine-containing agrichemicals are measurable in farmland runoff and nearby rivers.[71]

### 1.9 Fluoride In India

Professor A.K. Susheela in India provides tests and a fluoride intake assessment on patients at her Fluorosis Research and Rural Development clinic.

In India, the national policy on fluoride in drinking water is "the less fluoride the better, as fluoride is injurious to health". There is no silly argument saying that only drinking water with more than 1.5 (World Health Org. limit) or 4 ppm (U.S. limit) can cause illness and drinking water that has less than that can't possibly make you ill or affect your kidneys.

In India, if you have symptoms of fluoride poisoning, you get these tests. They are done without question regardless of how much or little fluoride is in the water you are drinking. If you have elevated fluoride in urine or blood, you've got fluoride poisoning.

In India there is nothing political to stop the diagnosis from being made or the patient properly treated, unlike here. (For a peek at the situation in India, see this Interview with Dr. Susheela.)

**1.10 References:**


1. Jaccaud, M.; Faron, R.; Devilliers, D.; Romano, R. (2000). Ullmann, Franz, ed. "Fluorine". *Ullmann's Encyclopedia of Industrial Chemistry* (Weinheim: Wiley-VCH) **15**: 381–395. doi:10.1002/14356007.a11_293. ISBN 3527306730.
2. Dean, John A. (1999). *Lange's Handbook of Chemistry* (15th ed.). New York: McGraw-Hill. ISBN 0-07-016190-9.
3. Lide, David R. (2004). *Handbook of Chemistry and Physics* (84th ed.). Boca Raton: CRC Press. ISBN 0-8493-0566-7.
4. Moore, John W.; Stanitski, Conrad L.; Jurs, Peter C. (2010). *Principles of Chemistry: The Molecular Science*. Belmont: Brooks/Cole. ISBN 978-0-495-39079-4.
5. Cordero, B.; Gómez, V.; Platero-Prats, A. E.; Revés, M.; Echeverría, J.; Cremades, E.; Barragán, F.; Alvarez, S. (2008). "Covalent Radii Revisited". *Dalton Transactions* (21): 2832–2838. doi:10.1039/b801115j.
6. Pyykkö, P.; Atsumi, M. (2009). "Molecular Double-Bond Covalent Radii for Elements Li–E112". *Chemistry - A European Journal* **15** (46): 12770–12779. doi:10.1002/chem.200901472.
7. William R. Dolbier, Jr. (2005). "Fluorine Chemistry at the Millennium". *Journal of Fluorine Chemistry* **126** (2): 157. doi:10.1016/j.jfluchem. 2004.09.033.
8. Roy J. Plunkett Chemical Heritage Foundation. Retrieved 10 September 2006.
9. G. Siegemund, W. Schwertfeger, A. Feiring, B. Smart, F. Behr, H. Vogel, B. McKusick "Fluorine Compounds, Organic" in "Ullmann's Encyclopedia of Industrial Chemistry" 2005, Wiley-VCH, Weinheim. doi:10.1002/14356007. a11_349.

10. J. H. Simons "The Electrochemical Process for the Production of Fluorocarbons" Journal of The Electrochemical Society, 1949, Volume 95, pp. 47-66. doi: 10.1149/ 1.2776733.
11. C. Heidelberger, N. K. Chaudhuri, P. Danneberg, D. Mooren, L. Griesbach, R. Duschinsky, R. J. Schnitzer, E. Pleven, and J. Schreiner (1957). "Fluorinated Pyrimidines, A New Class of Tumour-Inhibitory Compounds". *Nature* **179** (4561): 663. Bibcode:1957Natur.179..663H. doi:10.1038/179663a0. PMID 13418758.
12. O'Hagan, D; Schaffrath, C; Cobb, S. L; Hamilton, J. T; Murphy, C. D (2002). "Biochemistry: biosynthesis of an organofluorine molecule". *Nature.* **416** (6878): 279. Bibcode:2002Natur.416..279O. doi:10.1038/416279a. PMID 11907567.
13. Clayton 2003, pp. 101–104. Clayton, Donald (2003). *Handbook of Isotopes in the Cosmos: Hydrogen to Gallium.* New York: Cambridge University Press. ISBN 978-0-521-82381-4.
14. Renda et al. 2004. Renda, A.; Fenner, Y.; Gibson, B. K.; Karakas, A. I.; Lattanzio, J. C.; Campbell, S.; Chieffi, A.; Cunha, K.; Smith, V. V. (2004). "On the Origin of Fluorine in the Milky Way". *Monthly Notices of the Royal Astronomical Society* **354** (2): 575–581. arXiv:astro-ph/0410580. Bibcode:2004MNRAS.354..575R. doi:10. 1111/j.1365-2966.2004.08215.x.
15. Jaccaud et al. 2000, p. 384. Jaccaud, M.; Faron, R.; Devilliers, D.; Romano, R. (2000). Ullmann, Franz, ed. "Fluorine". *Ullmann's Encyclopedia of Industrial Chemistry* (Weinheim: Wiley-VCH) **15**: 381–395. doi:10.1002/14356007.a11_293. ISBN 3527306730.
16. Schulze-Makuch & Irwin 2008, p. 121. Schulze-Makuch, D.; Irwin, L. N. (2008). *Life in the Universe: Expectations and Constraints* (2nd ed.). Berlin: Springer-Verlag. ISBN 978-3-540-76816-6.
17. Haxel, Hedrick & Orris 2005. Haxel, G. B.; Hedrick, J. B.; Orris, G. J. (2005). □Rare Earth Elements—Critical Resources for High Technology, Fact Sheet 087-02□ (Report). U.S. Geological Survey. http://pubs.usgs.gov/fs/2002/fs087-02/. Retrieved 31 January 2014.
18. Greenwood & Earnshaw 1998, p. 795. Greenwood, N. N.; Earnshaw, A. (1998). *Chemistry of the Elements* (2nd ed.). Oxford: Butterworth Heinemann. ISBN 0-7506-3365-4.
19. Norwood & Fohs 1907, p. 52. Norwood, Charles J.; Fohs, F. Julius (1907). *Kentucky Geological Survey, Bulletin No. 9: Fluorspar Deposits of Kentucky.* Kentucky Geological Survey.

20. Villalba, Ayres & Schroder 2008. Villalba, G.; Ayres, R. U.; Schroder, H. (2008). "Accounting for Fluorine: Production, Use, and Loss". *Journal of Industrial Ecology* **11**: 85–101. doi:10.1162/jiec.2007.1075.
21. Kelly & Miller 2005. Kelly, T. D.; Miller, M. M. (2005). "Historical Fluorspar Statistics". U.S. Geological Service. Retrieved 10 February 2014.
22. Lusty et al. 2008. Lusty, P. A. J.; Brown, T. J.; Ward, J.; Bloomfield, S. (2008). "The Need for Indigenous Fluorspar Production in England". British Geological Survey. Retrieved 13 October 2013.
23. Gribble 2002. Gribble, G. W. (2002). Neison, A. H., ed. "Naturally Occurring Organofluorines". *Organofluorines* (Berlin: Springer): 121–136. doi:10.1007/107-21878_5. ISBN 3-540-42064-9.
24. Richter, Hahn & Fuchs 2001, p. 3. Richter, M.; Hahn, O.; Fuchs, R. (2001). "Purple Fluorite: A Little Known Artists' Pigment and Its Use in Late Gothic and Early Renaissance Painting in Northern Europe". *Studies in Conservation* **46** (1): 1–13. doi:10.1179/sic.2001.46.1.1. JSTOR 1506878.
25. Schmedt, Mangstl & Kraus 2012. Schmedt, J.; Mangstl, M.; Kraus, F. (2012). "Occurrence of Difluorine $F_2$ in Nature—In Situ Proof and Quantification by NMR Spectroscopy". *Angewandte Chemie International Edition* **51** (31): 7847–7849. doi:10.1002/ange.201203515.
26. Senning 2007, p. 149. Senning, A. (2007). *Elsevier's Dictionary of Chemoetymology: The Whies and Whences of Chemical Nomenclature and Terminology*. Amsterdam and Oxford: Elsevier. ISBN 978-0-444-52239-9.
27. Stillman 1912. Stillman, John Maxson (December 1912). "Basil Valentine, A Seventeenth Century Hoax". *Popular Science Monthly* **81**. Retrieved 14 October 2013.
28. Gribble, Gordon W. (2002). "Naturally occurring organofluorines". *The Handbook of Environmental Chemistry*. The Handbook of Environmental Chemistry **3N**: 121–136. doi:10.1007/10721878_5. ISBN 3-540-42064-9.
29. Murphy, C.; Schaffrath, C.; O'Hagan, D. (2003). "Fluorinated natural products: The biosynthesis of fluoroacetate and 4-fluorothreonine in *Streptomyces cattleya*". *Chemosphere* **52** (2): 455–461. doi:10.1016/S0045-6535(03)00191-7.
30. Proudfoot, A. T.; Bradberry, S. M.; Vale, J. A. (2006). "Sodium fluoroacetate poisoning". *Toxicological Reviews* **25** (4): 213–219. doi:10.2165/001397092006-25040-00002. PMID 17288493.

31. O'Hagan, D.; Schaffrath, C.; Cobb, S. L.; Hamilton, J. T.; Murphy, C. D. (2002). "Biochemistry: Biosynthesis of an organofluorine molecule". *Nature* **416** (6878): 279. Bibcode:2002Natur.416..279O. doi:10.1038/ 416279a. PMID 11907567.
32. Olivares, M.; Uauy, R. (2004). □Essential nutrients in drinking water (Draft)□ (Report). WHO Retrieve 30 December 2008. http://www.who.int/water_sanitation_health/dwq/en/nutoverview.pdf.
33. Nielsen, Forrest H. (2009). "Micronutrients in parenteral nutrition: Boron, silicon, and fluoride". *Gastroenterology* **137** (5 Suppl): S55–S60. doi: 10.1053/j.gastro. 2009.07. 072. PMID 19874950.
34. Greenwood & Earnshaw 1998, p. 804. Greenwood, N. N.; Earnshaw, A. (1998). *Chemistry of the Elements* (2nd ed.). Oxford: Butterworth Heinemann. ISBN 0-7506-3365-4.
35. Macomber 1996, p. 230 Macomber, Roger (1996). *Organic chemistry* **1**. Sausalito: University Science Books. ISBN 978-0-935702-90-3.
36. Jaccaud et al. 2000, p. 382. Jaccaud, M.; Faron, R.; Devilliers, D.; Romano, R. (2000). Ullmann, Franz, ed. "Fluorine". *Ullmann's Encyclopedia of Industrial Chemistry* (Weinheim: Wiley-VCH) **15**: 381–395. doi:10.1002/14356007.a11_293. ISBN 3527306730.
37. Nelson 1947. Nelson, Eugene W. (1947). "'Bad Man' of The Elements". *Popular Mechanics* **88** (2): 106–108, 260.
38. Lidin, Molochko & Andreeva 2000, pp. 442–455. Lidin, R.; Molochko, V.A.; Andreeva, L.L. (2000). *Химические свойства неорганических веществ* [*Chemical Properties of Inorganic Substances*] (in Russian). Moscow: Khimiya. ISBN 5-7245-1163-0.
39. Wiberg, Wiberg & Holleman 2001, p. 404. Wiberg, Egon; Wiberg, Nils; Holleman, Arnold Frederick (2001). *Inorganic Chemistry*. San Diego: Academic Press. ISBN 978-0-12-352651-9.
40. Patnaik 2007, p. 472. Patnaik, Pradyot (2007). *A Comprehensive Guide to the Hazardous Properties of Chemical Substances* (3rd ed.). Hoboken: John Wiley & Sons. ISBN 978-0-471-71458-3.
41. Aigueperse et al. 2000, p. 400. Aigueperse, J.; Mollard, P.; Devilliers, D.; Chemla, M.; Faron, R.; Romano, R. E.; Cue, J. P. (2000). Ullmann, Franz, ed. "Fluorine Compounds, Inorganic". *Ullmann's Encyclopedia of Industrial Chemistry* (Weinheim: Wiley-VCH) **15**: 397–441. doi:10.1002/14356007. ISBN 3527306730.

42. Greenwood & Earnshaw 1998, pp. 76, 804. Greenwood, N. N.; Earnshaw, A. (1998). *Chemistry of the Elements* (2nd ed.). Oxford: Butterworth Heinemann. ISBN 0-7506-3365-4.
43. Kuriakose & Margrave 1965. Kuriakose, A. K.; Margrave, J. L. (1965). "Kinetics of the Reactions of Elemental Fluorine. IV. Fluorination of Graphite". *Journal of Physical Chemistry* **69** (8): 2772–2775. doi:10.1021/j100892a049.
44. Hasegawa et al. 2007. Hasegawa, Y.; Otani, R.; Yonezawa, S.; Takashima, M. (2007). "Reaction Between Carbon Dioxide and Elementary Fluorine". *Journal of Fluorine Chemistry* **128** (1): 17–28. doi: 10.1016/j.jfluchem.2006.09.002.
45. Lagow 1970, pp. 64–78. Lagow, R. J. (1970). *The Reactions of Elemental Fluorine; A New Approach to Fluorine Chemistry* (PhD thesis, Rice University, TX). Ann Arbor: UMI.
46. Navarrini et al. 2012. Navarrini, W.; Venturini, F.; Tortelli, V.; Basak, S.; Pimparkar, K. P.; Adamo, A.; Jensen, K. F. (2012). "Direct Fluorination of Carbon Monoxide in Microreactors". *Journal of Fluorine Chemistry* **142**: 19–23. doi:10.1016/j.jfluchem. 2012.06.006.
47. Lidin, Molochko & Andreeva 2000, p. 252. Lidin, R.; Molochko, V.A.; Andreeva, L.L. (2000). *Химические свойства неорганических веществ* [*Chemical Properties of Inorganic Substances*] (in Russian). Moscow: Khimiya. ISBN 5-7245-1163-0.
48. Tanner Industries 2011. Tanner Industries (January 2011). "Anhydrous Ammonia: (MSDS) Material Safety Data Sheet". tannerind.com. Retrieved 24 October 2013.
49. Morrow, Perry & Cohen 1959. Morrow, S. I.; Perry, D. D.; Cohen, M. S. (1959). "The Formation of Dinitrogen Tetrafluoride in the Reaction of Fluorine and Ammonia". *Journal of the American Chemical Society* **81** (23): 6338–6339. doi:10.1021/ja01532-a066.
50. Emeléus & Sharpe 1974, p. 111. Emeléus, H. J.; Sharpe, A. G. (1974). *Advances in Inorganic Chemistry and Radiochemistry* **16**. New York: Academic Press. ISBN 978-0-08-057865-1.
51. Wiberg, Wiberg & Holleman 2001, p. 457. Wiberg, Egon; Wiberg, Nils; Holleman, Arnold Frederick (2001). *Inorganic Chemistry*. San Diego: Academic Press. ISBN 978-0-12-352651-9.
52. Brantley 1949, p. 26. Brantley, L. R. (1949). Squires, Roy; Clarke, Arthur C., eds. "Fluorine". *Pacific Rockets: Journal of the Pacific Rocket Society* (South Pasadena: Sawyer Publishing/Pacific Rocket Society Historical Library) **3** (1): 11–18. ISBN 978-0-9794418-5-1.

53. Jaccaud et al. 2000, p. 383. Jaccaud, M.; Faron, R.; Devilliers, D.; Romano, R. (2000). Ullmann, Franz, ed. "Fluorine". *Ullmann's Encyclopedia of Industrial Chemistry* (Weinheim: Wiley-VCH) **15**: 381–395. doi:10.1002/14356007.a11_293. ISBN 3527306730.
54. Pitzer 1975. Pitzer, K. S. (1975). "Fluorides of Radon and Element 118". *Journal of the Chemical Society, Chemical Communications* (18): 760b–761. doi:10.1039/C3975000760B.
55. Khriachtchev et al. 2000. Khriachtchev, L.; Pettersson, M.; Runeberg, N.; Lundell, J.; Räsänen, M. (2000). "A Stable Argon Compound". *Nature* **406** (6798): 874–876. doi:10.1038/35022551. PMID 10972285 Greenwood & Earnshaw 1998, p. 804. Greenwood, N. N.; Earnshaw, A. (1998). *Chemistry of the Elements* (2nd ed.). Oxford: Butterworth Heinemann. ISBN 0-7506-3365-4.
56. Barry & Phillips 2006. Barry, Patrick L.; Phillips, Tony (26 May 2006). "Good News and a Puzzle". National Aeronautics and Space Administration. Retrieved 6 January 2012.
57. EPA 2013a.
58. EPA 2013b.
59. McCoy 2007. McCoy, M. (2007). "SURVEY Market Challenges Dim the Confidence of the World's Chemical CEOs". *Chemical & Engineering News* **85** (23): 11. doi:10.1021/cen-v085n023.p011a.
60. Forster et al. 2007, pp. 212–213. Forster, P.; Ramaswamy, V; Artaxo, P.; Berntsen, T.; Betts, R.; Fahey, D. W.; Haywood, J.; Lean, J.; Lowe, D. C.; Myhre, G.; Nganga, J.; Prinn, R.; Raga, G.; Schulz, M.; Van Dorland, R. (2007). "Changes in Atmospheric Constituents and in Radiative Forcing". In Solomon, S.; Manning, M; Chen, Z.; Marquis, M; Averyt, K. B.; Tignor, M.; Miller, H. L. (eds.). *Climate Change 2007: The Physical Science Basis. Contribution of Working Group I to the Fourth Assessment Report of the Intergovernmental Panel on Climate Change.* Cambridge: Cambridge University. pp. 129–234. ISBN 978-0-521-70596-7.
61. Schwarcz 2004, p. 37. Schwarcz, Joseph A. (2004). *The Fly in the Ointment: 70 Fascinating Commentaries on the Science of Everyday Life*. Toronto: ECW Press. ISBN 1-55022-621-5.
62. Giesy & Kannan 2002. Giesy, J.P.; Kannan, K. (2002). "Perfluorochemical Surfactants in the Environment". *Environmental Science & Technology* **36** (7): 146A–152A. doi:10.1021/es022253t. PMID 11999053.

63. Buznik, V. M. (2009). "Fluoropolymer Chemistry in Russia: Current Situation and Prospects". *Russian Journal of General Chemistry* **79** (3): 520–526. doi:10.1134/S1070363209030335.
64. http://pubs.acs.org/doi/pdf/10.1021/es022253t
65. Steenland, Fletcher & Savitz 2010. Steenland, K.; Fletcher, T.; Savitz, D. A. (2010). "Epidemiologic Evidence on the Health Effects of Perfluorooctanoic Acid (PFOA)". *Environmental Health Perspectives* **118** (8): 1100–1108. doi:10.1289/ ehp.0901827. PMC 2920088. PMID 20423814.
66. Betts 2007. Betts, K. S. (2007). "Perfluoroalkyl Acids: What is the Evidence Telling Us?". *Environmental Health Perspectives* **115** (5): A250–A256. doi:10.1289/ehp.115-a250. PMC 1867999. PMID 17520044.
67. http://www.epa.gov/fedfac/pdf/emerging_contaminants_pfos_pfoa.pdf
68. Perfluorooctanoic acid (PFOA) and perfluorooctanesulfonic acid (PFOS) in surface waters, sediments, soils and wastewater – A review on concentrations and distribution coefficients.
69. http://toxsci.oxfordjournals.org/content/99/2/366.full.pdf
70. Lietz & Meyer 2006, pp. 7–8. Lietz, A. C.; Meyer, Michael T. (2006). □Evaluation of Emerging Contaminants of Concern at the South District Waste Water Treatment Plant Based on Seasonal Sampling Events, Miami-Dade Country, Florida, 2004□ (Report). U.S. Geological Survey Scientific Investigations. http://pubs.usgs.gov/sir/2006/5240/pdf/sir 2006-5240.pdf. Retrieved 6 June 2011.
71. Ahrens 2011. Ahrens, L. (2011). "Polyfluoroalkyl Compounds in the Aquatic Environment: A Review of Their Occurrence and Fate". *Journal of Environmental Monitoring* (Royal Society of Chemistry) **13** (1): 20–31. doi:10.1039/c0em00373e. PMID 21031178.Macomber 1996, p. 230 Macomber, Roger (1996). *Organic chemistry* **1**. Sausalito: University Science Books. ISBN 978-0-935702-90-3.

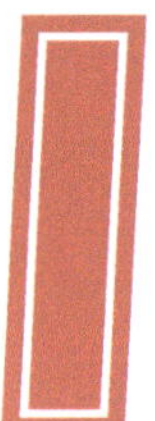

# Chapter-2
# Fluorine Applications

## 2. Fluorine Applications

Organofluorine chemistry impacts many areas of everyday life and technology. The C-F bond is found in pharmaceuticals, agrichemicals, fluoropolymers, refrigerants, surfactants, anesthetics, oil-repellents, catalysis, and water-repellents, among others.

### 2.1 Pharmaceuticals:

The carbon-fluorine bond is commonly found in pharmaceuticals because it is generally metabolically stable and fluorine acts as a bioisostere of the hydrogen atom. All commercialized pharmaceutical drugs, 20% contain fluorine, including important drugs in many different pharmaceutical classes.[1] Fluorine is added to drug molecules as even a single atom of it can greatly change the chemical properties of the molecule in ways that are desirable.

Because of the considerable stability of the carbon-fluorine bond, many drugs are fluorinated to delay their metabolism, which is the chemical process in which the drugs are turned into compounds that allows them to be eliminated. This prolongs their half-lives and allows for longer times between dosing and activation. For example, an aromatic ring may prevent the metabolism of a drug, but this presents a safety problem, because some aromatic compounds are metabolized in the body into poisonous epoxides by the organism's native enzymes. Substituting fluorine into a *para* position, however, protects the aromatic ring and prevents the epoxide from being produced.

Adding fluorine to biologically active organics increases their lipophilicity (ability to dissolve in fats), because the carbon–fluorine bond is even more hydrophobic than the carbon–hydrogen bond. This effect often increases a drug's bioavailability because of increased cell membrane penetration.[2] Although the potential of fluorine being released in a fluoride leaving group depends on its position in the molecule,[3] organofluorides are generally very stable, since the carbon–fluorine bond is strong.

Fluorines also find their uses in common mineralocorticoids, a class of drugs that increase the blood pressure. Adding fluorine increases both its medical power and anti-inflammatory effects.[4] Fluorine-containing fludrocortisone is one of the most common of these drugs.[5] Dexamethasone and triamcinolone, which are among the most potent of the related synthetic corticosteroids class of drugs, contain fluorine as well.[5]

Several inhaled general anesthetic agents, including the most commonly used inhaled agents, also contain fluorine. The first fluorinated anesthetic agent, halothane, proved to be much safer (neither explosive nor flammable) and longer-lasting than those previously used. Modern fluorinated anesthetics are longer-lasting still and almost insoluble in blood, which accelerates the awakening.[6] Examples include sevoflurane, desflurane, enflurane, and isoflurane, all hydrofluorocarbon derivatives.[7]

Prior to 1980s, antidepressants altered not only the serotonin uptake (lack of serotonin is a reason for a depression), but also the altered norepinephrine uptake; this caused most antidepressants' side effects. The first drug to only alter the serotonin uptake was Prozac; it gave birth to the extensive selective serotonin reuptake inhibitor (SSRI) antidepressants class and is the best-selling antidepressant. Many other SSRI antidepressants are fluorinated organics, including Celexa, Luvox, and Lexapro.[8] Fluoroquinolones are a commonly used family of broad-spectrum antibiotics.[9]

An estimated one fifth of pharmaceuticals contain fluorine, including several of the top drugs.[10]

- 2. Anticancer Drugs
  - 2.1. Fulvestrant (Faslodex)
  - 2.2. Gefitinib (Iressa)
  - 2.3. Sorafenib (Nexavar)
  - 2.4. Capecitabine (Xeloda)
  - 2.5. Sunitinib (Sutent)
  - 2.6. Nilotinib (Tasigna)
  - 2.7. Lapatinib (Tykerb)
  - 2.8. Crizotinib (Xalkori)
  - 2.9. Vemurafenib (Zelboraf)
  - 2.10. Vandetanib (Caprelsa)
- 3. Drugs Acting on the Central Nervous System
  - 3.1. Escitalopram (Lexapro)
  - 3.2. Aprepitant (Emend)

- 9. Anti-Diabetes Drugs
  - 9.1. Sitagliptin (Januvia, Janumet)
- 10. Gastrointestinal Tract Drugs
  - 10.1. Lubiprostone (Amitiza)
- 11. Endocrine System Drugs
  - 11.1. Cinacalcet (Sensipar, Minpara)
- 12. Nutrition Affecting Drugs
  - 12.1. Nitisinone (Orfadin)

**2.2 Agrochemicals:**

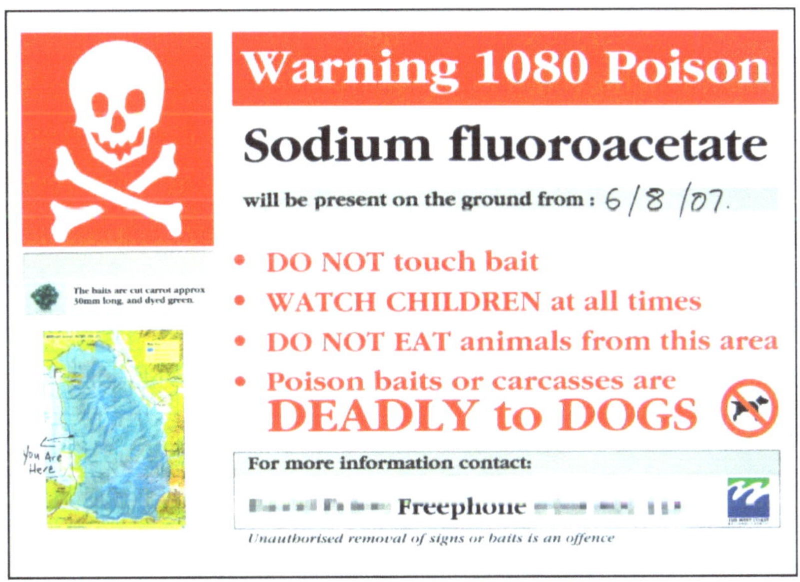


*Sign warning of poisonous sodium fluoroacetate baits*

Synthetic sodium fluoroacetate has been used as an insecticide but is especially effective against mammalian pests.[11] The name "1080" refers to the catalogue number of the poison, which became its brand name.[12] Fluoroacetate is similar to acetate, which has a pivotal role in the Krebs cycle (a key part of cell metabolism). Fluoroacetate halts the cycle and causes cells to be deprived of energy.[12] Several other insecticides contain sodium fluoride, which is much less toxic than fluoroacetate.[13]

An estimated 30% of agrichemical compounds contain fluorine.[14] Most of them are poisons, but a few stimulate the growth instead. It is expected that how often the fluorine agrichemicals will be used depends on two factors: if the synthesis reaction will be improved (to reduce the prices) and if green chemistry will be taken in account to a larger scale (fluorochemicals are more environment-friendly).[15]

An important agrichemcial is Trifluralin. It was once very important (for example, in 1998 over a half of U.S. cotton field area was coated with the chemical [16]); however, its suspected carcinogenic properties caused some Northern European countries to ban it in 1993.[17] Currently, the whole European Union has it banned, although there was a case intended to cancel the decision.[18]

The currently used agrichemicals utilize another tactic: instead of being poisonous themselves, e.g., by directly affecting metabolism, they transform the metabolism so the organism produces poisonous compounds. For example, insects fed 29-fluorostigmasterol produce the fluoroacetates from it. If fluorine is transferred to a body cell, it blocks metabolism at the position occupied.[19]

**Recent development in agrochemical:**

The past 15 years has been a very exciting time for research into agrochemicals with many significant advances being achieved. Highly active and effective new series of compounds have been discovered in each of the agrochemical disciplines (herbicides, insecticide, fungicides and plant growth regulator), and compound have been produced to the market which are often an order of magnitude more active than earlier products. This has all been achieved against a background of change in the agrochemical industry. Finney has presented figures to show that while there has been a progressive decline in the profitability of the industry, expenditure on research and development has increased very substantially.1 this reflects the substantial increase in development costs required to satisfy increasing regulatory demands. Graham-Bryce has argued that these demands need to be considered in their proper perspective.2 fluorine as a substituent has played a significant and increasingly important part in the development of new agrochemicals and is likely to continue to do so in the future.

This review concentrates on compounds introduced to the market during the last five years, i.e., since Newbold's review.3 examples have been selected to demonstrate the variety of fluorine substituents and substitution patterns used in agrochemicals, and to show how they are assembled from key building blocks. Since fluorine-containing molecules are generally more expensive than nonfluorinated analogues, a clear advantage needs to be gained to justify the inclusion of fluorine. The required fluorine-containing starting materials are obtained where possible form

**2.3 Inhaler Propellant**:

This is also used as a propellant for inhalers used to administer asthma medications. It is referred to as HFA and replaces CFC propellant based inhalers. CFC inhalers were banned as of 2008 because of environmental concerns with the ozone layer. HFA propellant inhalers like FloVent and ProAir (Salbutamol) have no generic versions available as of November 2012.

**2.4 Fluorosurfactants**:

Fluorosurfactants, which have a polyfluorinated "tail" and a hydrophilic "head", serve surfactants because they concentrate at the liquid-air interface due to their lipophobicity. Fluorosurfactants have low surface energies and dramatically lower surface tension. The fluorosurfactants perfluorooctanesulfonic acid (PFOS) and perfluorooctanoic acid (PFOA) are two of the most studied because of their ubiquity, toxicity, and long residence times in humans and wildlife.

**2.5 Solvents**:

Fluorinated compounds often display distinct solubility properties. Dichlorodifluoromethane and chlorodifluoromethane were widely used refrigerants. CFCs have potent ozone depletion potential due to the homolytic cleavage of the carbon-chlorine bonds; their use is largely prohibited by the Montreal Protocol. Hydrofluorocarbons (HFCs), such as tetrafluoroethane, serve as CFC replacements because they do not catalyze ozone depletion. Oxygen exhibits a high solubility in perfluorocarbon compounds, reflecting again on their lipophilicity. Perfluorodecalin has been demonstrated as a blood substitutes, transporting oxygen to the lungs.

The solvent 1,1,1,2-tetrafluoroethane has been used for extraction of natural products such as taxol, evening primrose oil, and vanillin. 2,2,2-trifluoroethanol is an oxidation-resistant polar solvent.[20]

**2.6 Scanning**:

Main articles: Positron emission tomography and Fludeoxyglucose (18F)

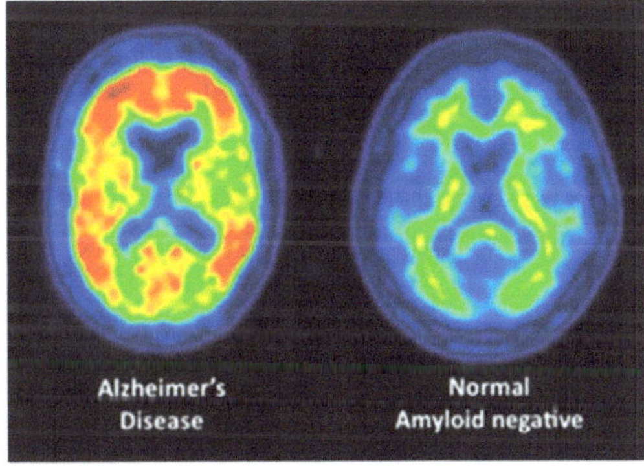


*PET scan for Alzheimer's diagnosis*

Compounds containing fluorine-18, a radioactive isotope that emits positrons, are often used in positron emission tomography (PET) scanning, because the isotope's half-life of about 110 minutes is long by positron-emitter standards. One such radiopharmaceutical is 2-deoxy-2-($^{18}$F)fluoro-D-glucose (generically referred to as fludeoxyglucose), commonly abbreviated as $^{18}$F-FDG, or simply FDG.[21] In PET imaging, FDG can be used for assessing glucose metabolism in the brain and for imaging cancer tumors. After injection into the blood, FDG is taken up by "FDG-avid" tissues with a high need for glucose, such as the brain and most types of malignant tumors.[22] Tomography, often assisted by a computer to form a PET/CT (CT stands for "computer tomography") machine, can then be used to diagnose or monitor treatment of cancers; especially Hodgkin's lymphoma, lung cancer, and breast cancer.[23]

Natural fluorine is monoisotopic, consisting solely of fluorine-19. Fluorine compounds are highly amenable to nuclear magnetic resonance (NMR), because fluorine-19 has a nuclear spin of ½, a high nuclear magnetic moment, and a high magnetogyric ratio. Fluorine compounds typically have a fast NMR relaxation, which enables the use of fast averaging to obtain a signal-to-noise ratio similar to hydrogen-1 NMR spectra.[24] Fluorine-19 is commonly used in NMR study of metabolism, protein structures and conformational changes.[25] In addition, inert fluorinated gases have the potential to be a cheap and efficient tool for imaging lung ventilation.[26]

## 2.7 Oxygen transport research:

See also: Blood substitute and Liquid breathing

Liquid fluorocarbons have a very high capacity for holding gas in solution. They can hold more oxygen or carbon dioxide than blood does. For that reason, they have attracted ongoing interest related to the possibility of artificial blood or of liquid breathing.[27]

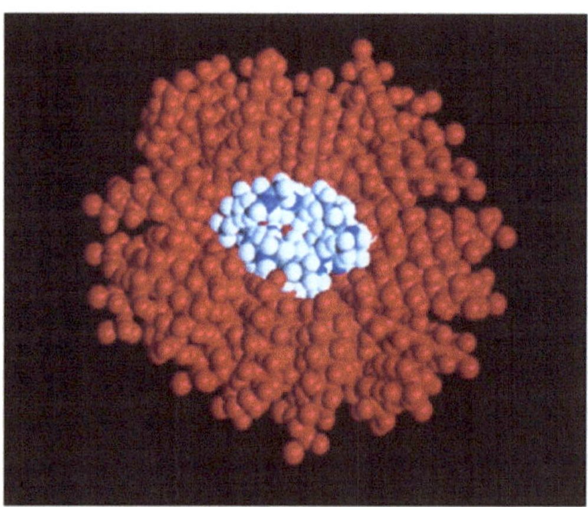

*Computer-generated model of nanocrystal of perflubron (red) and gentamicin (white, an antibiotic)*

Blood substitutes are the subject of research because the demand for blood transfusions grows faster than donations. In some scenarios, artificial blood may be more convenient or safe. Because fluorocarbons do not normally mix with water, they must be mixed into emulsions (small droplets of perfluorocarbon suspended in water) to be used as blood.[28][29] One such product, Oxycyte, has been through initial clinical trials.[30][31]

Possible medical uses of liquid breathing (which uses pure perfluorocarbon liquid, not a water emulsion) involve assistance for premature babies or for burn victims (because the normal lung function is compromised). Both partial filling of the lungs and complete filling of the lungs have been considered, although only the former has any significant tests in humans. Several animal tests have been performed and some human partial liquid ventilation trials.[32] One effort, by Alliance Pharmaceuticals, reached clinical trials but was abandoned because of insufficient advantage compared to other therapies.[33]

Nanocrystals represent a possible method of delivering water or fat soluble drugs within a perfluorochemical fluid. The particles usage is being developed to help treat babies with damaged lungs.[34]

Perfuorocarbons are banned from sports, where they may be used to increase oxygen use for endurance athletes. One cyclist, Mauro Gianetti was investigated after a near fatality where PFC use was suspected.[35][36]

Other posited applications include deep sea diving and space travel, applications that would both require total liquid ventilation, not partial ventilation.[37][38] The 1989 film The Abyss showed a fictional use of perfluorocarbon for human diving but also filmed a real rat surviving while cooled and immersed in perfluorocarbon.[39] (See also list of fictional treatments of perfluorocarbon breathing.)

**2.8 Industrial:**

The largest application of fluorine, consuming up to 7,000 metric tons annually, is in the preparation of $UF_6$ for the nuclear fuel cycle. Fluorine is used to fluorinate uranium tetrafluoride, itself formed from uranium dioxide and hydrofluoric acid. Fluorine is monoisotopic, so any mass differences between $UF_6$ molecules are due to the presence of 235U or 238U, enabling uranium enrichment via diffusion or centrifuge. About 6,000 metric tons per year go into producing the inert dielectric SF6 for high-voltage transformers and circuit breakers, eliminating the need for hazardous polychlorinated biphenyls associated with oil-filled devices. Several fluorine compounds are used in electronics: rhenium and tungsten hexafluoride in chemical vapor deposition, tetra-fluoromethane in plasma etching and nitrogen trifluoride in cleaning equipment. Fluorine is also used in the synthesis of organic fluorides, but its reactivity often necessitates conversion first to the gentler $ClF_3$, $BrF_3$, or $IF_5$, which together allow calibrated fluorination. Fluorinated pharmaceuticals use sulfur tetrafluoride instead.

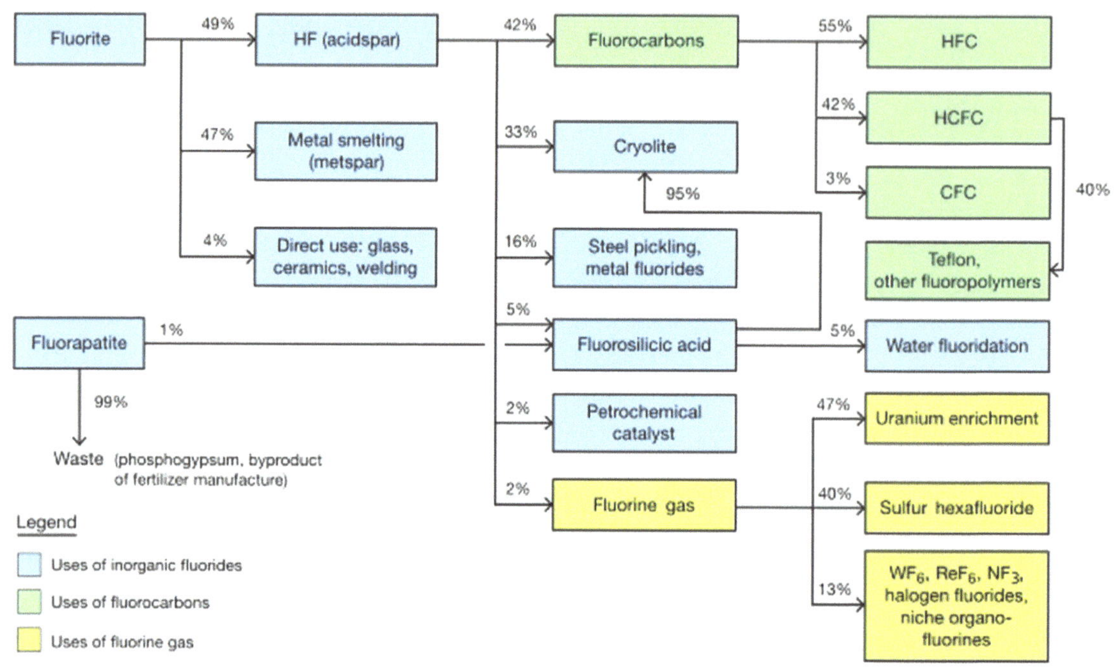


## 2.9 References:


1. Emsley, John (2011). *Nature's building blocks: An A–Z guide to the elements* (2nd ed.). Oxford University Press. p. 178. ISBN 978-0-19-960563-7.
2. Swinson, Joel (2005). "Fluorine – A vital element in the medicine chest". *PharmaChem* (Pharmaceutical Chemistry): 26–27. Retrieved 26 August 2010.
3. Schubiger, P. A. (2006). *Pet chemistry: The driving force in molecular imaging.* Springer. p. 144. ISBN 9783540326236.
4. Goulding, Nicolas J.; Flower, Rod J. (2001). *Glucocorticoids*. Springer. p. 40. ISBN 9783764360597.
5. Raj, P. Prithvi; Erdine, Serdar (2012). *Pain-relieving procedures: The illustrated guide*. John Wiley & Sons. p. 58. ISBN 9781118300459.
6. Bégué, Jean-Pierre; Bonnet-Delpon, Daniele (2008). *Bioorganic and Medicinal Chemistry of Fluorine*. John Wiley & Sons. pp. 335–336. ISBN 9780470281871.
7. Filler, R.; Saha, R. (2009). "Fluorine in medicinal chemistry: A century of progress and a 60-year retrospective of selected highlights". *Future Medicinal Chemistry* **1** (5): 777–791. doi:10.4155/fmc.09.65. PMID 21426080.
8. Mitchell, E. Siobhan; Triggle, D. J. (2004). *Antidepressants*. Infobase Publishing. pp. 37–39. ISBN 978-1-4381-0192-7.
9. Nelson, J. M.; Chiller, T. M.; Powers, J. H.; Angulo, F. J. (2007). "Fluoroquinolone resistant Campylobacter species and the withdrawal of fluoroquinolones from use

in poultry: a public health success story". *Clinical Infectious Diseases* **44** (7): 977–980. doi:10.1086/512369. PMID 17342653.

10. Ann M. Thayer "Fabulous Fluorine" Chemical and Engineering News, June 5, 2006, Volume 84, pp. 15-24. http://pubs.acs.org/cen/coverstory/84/8423cover1.html
11. Eisler, Ronald (1995). □Biological report 27: Sodium monofluoroacetate (1080) Hazards to fish, wildlife and invertebrates: A synoptic review□ (Report). Patuxent Environmental Science Center (U.S. National Biological Service). http://www.pwrc.usgs.gov/infobase/eisler/CHR_30_Sodium_monofluoroacetate.pdf. Retrieved 5 June 2011.
12. Proudfoot, A. T.; Bradberry, S. M.; Vale, J. A. (2006). "Sodium fluoroacetate poisoning". *Toxicological Reviews* **25** (4): 213–219. doi: 10.2165/00139709-200625040-00002. PMID 17288493.
13. "Class I ozone-depleting substances". *Sodium fluoride – pesticidal uses*. Scorecard. Retrieved 20 February 2011.
14. "Fluorine's treasure trove". ICIS news. 2006-10-02. Retrieved 20 February 2011.
15. Theodoridis, George (2006). *Fluorine and the Environment: Agrochemicals, Archaeology, Green Chemistry & Water*. Elseiver. pp. 121–176. ISBN 9780444526724.
16. "Fact sheet: Trifluralin". *Pesticides News* **52**: 20–21. 2001.
17. European Commission (2007). □Trifluralin□ (Report). http://www.unece.org/fileadmin/DAM/env/lrtap/TaskForce/popsxg/2008/Trifluralin_RA%20dossier_proposal%20for%20submission%20to%20the%20UNECE%20POP%20Protocol.pdf.
18. Case T-475/07, Dow AgroSciences Ltd vs. European Commission (2011). The General Court of European Union (Third Camber).
19. Barnette, William E. (1995). "Physical Organic Aspects of Fluorinated Argichemicals". *Fluorine in agriculture*. Smithers Rapra Publishing. pp. 1–19. ISBN 9781859570333.
20. Kabayadi S. Ravikumar, Venkitasamy Kesavan, Benoit Crousse, Danièle Bonnet-Delpon, and Jean-Pierre Bégué (2003), "Mild and Selective Oxidation of Sulfur Compounds in Trifluorethanol: Diphenyl Disulfide and Methyle Phenyl Sulfoxide", *Org. Synth.* **80**: 184.
21. Schmitz, A.; Kälicke, T.; Willkomm, P.; Grünwald, F.; Kandyba, J.; Schmitt, O. (2000). "Use of fluorine-18 fluoro-2-deoxy-D-glucose positron emission tomography in assessing the process of tuberculous spondylitis". *Journal of spinal*

*disorders* **13** (6): 541–544. doi:10.1097/00002517-200012000-00016. PMID 11132989.

22. Bustamante, Ernesto; Pedersen, Peter L. (1977). "High aerobic glycolysis of rat hepatoma cells in culture: Role of mitochondrial hexokinase". *Proceedings of the National Academy of Sciences of the United States of America* **74** (9): 3735–3739. Bibcode:1977PNAS...74.3735B. doi:10.1073/pnas.74.9.3735. PMC 431708. PMID 198801.
23. Hayat, M. A. (2007). *Cancer imaging: Lung and breast carcinomas*. Academic Press. p. 41. ISBN 9780123742124.
24. Nelson, J. H. (2003). *Nuclear magnetic resonance spectroscopy*. Prentice Hall. pp. 129–139. ISBN 0-13-033451-0.
25. Danielson, Mark A.; Falke, Joseph J. (1996). "Use of $^{19}F$ NMR to probe protein structure and conformational changes". *Annual Review of Biophysics and Biomolecular Structure* **25**: 163–195. doi:10.1146/ annurev.bb.25.060196.001115. PMC 2899692. PMID 8800468.
26. Kuethe, Dean O.; Caprihan, Arvind; Fukushima, Eiichi; Waggoner, R. Allen (2005). "Imaging lungs using inert fluorinated gases". *Magnetic Resonance in Medicine* **39** (1): 85–88. doi:10.1002/mrm.1910390114. PMID 9438441.
27. Gabriel, J. L.; Miller, T. F.; Wolfson, M. R. Jr; Shaffer, T. H. (1996). "Quantitative structure-activity relationships of perfluorinated hetro-hydrocarbons as potential respiratory media. Application to oxygen solubility, partition coefficient, viscosity, vapor pressure, and density". *ASAIO Journal* **42** (6): 968–973. doi:10.1097/00002480-199642060-00009. PMID 8959271.
28. Sarkar, S. (2008). "Artificial Blood". *Indian Journal of Critical Care Medicine* **12** (3): 140–144. doi:10.4103/0972-5229.43685. PMC 2738310. PMID 19742251.
29. Schimmeyer, S. (2002). "The search for a blood substitute". *Illumin* (University of Southern Carolina) **5** (1). Retrieved 2 December 2010.
30. Tasker, Fred (2008-03-19). *Miami Herald: Artificial blood goes from science fiction to science fact*. Miami Herald (at noblood.org). Archived from the original on 19 March 2008.
31. Davis, Nicole (2006). "Better than blood". *Popular Science*. Archived from the original on 2011-06-04. Retrieved 30 September 2012.
32. Shaffer, T. H.; Wolfson, M. R.; Clark, L. R. (1992). "State of art review: Liquid ventilation". *Pediatric Pulmonology* **14** (102–109): 102. doi: 10.1002/ppul.195014-0208. PMID 1437347.

33. Kacmarek, R. M.; Wiedemann, H. P.; Lavin, P. T.; Wedel, M. K.; Tütüncü, A. S.; Slutsky, A. S. (2006). "Partial Liquid Ventilation in Adult Patients with Acute Respiratory Distress Syndrome". *American Journal of Respiratory and Critical Care Medicine* **173** (8): 882–889. doi:10.1164/rccm.200508-1196OC. PMID 16254269.
34. Shaffer, Thomas H.; Wolfson, Marla R.; Greenspan, Jay S. (1999). "Liquid ventilation: Current status". *Pediatrics in Review* **20** (12): e134–e142. doi:10.1542/pir.20-12-e134. PMID 10587539.
35. Gains, Paul (October 18, 1998). "A New Threat in Blood Doping". *New York Times*.
36. http://www.salon.com/1999/04/21/cycling/
37. Kylstra, J. A. (1977). *The feasibility of liquid breathing in man.* Duke University. Retrieved 5 May 2008.
38. The Global Oneness Commitment. "Liquid breathing – Space travel". experiencefestival.com. Retrieved 17 May 2008.
39. Aljean Harmetz (1989). "FILM; 'The Abyss': A foray into deep waters". *The New York Times*. Retrieved 2 October 2012.
40. E.G. Hopea, A.P. Abbotta, D.L. Daviesa, G.A. Solana and A.M. Stuarta "Green Organometallic Chemistry" in Comprehensive Organometallic Chemistry III, 2007, Volume 12, Pages 837-864. doi:10.1016/B0-08-045047-4/00182-5
41. J. A. Gladysz, D. P. Curran, I. T. Horváth (Eds.) "Handbook of Fluorous Chemistry", Wiley VCH, Weinheim, 2004. ISBN 978-3-527-30617-6.
42. Aimee Crombie, Sun-Young Kim, Sabine Hadida, and Dennis P. Curran, "Synthesis of Tris(2-Perfluorohexylethyl)tin Hydride: A Highly Fluorinated Tin Hydride with Advantageous Features of Easy Purification", *Org. Synth.*; *Coll. Vol.* **10**: 712.
43. M.E. Thompson, P.E. Djurovich, S. Barlow and S. Marder "Organometallic Complexes for Optoelectronic Applications" Comprehensive Organometallic Chemistry III, 2007, Volume 12, Pages 101-194. doi:10.1016/B0-08-045047-4/00169-2.
44. J.C. Peters, J.C. Thomas "Ligands, Reagents, and Methods in Organometallic Synthesis" in Comprehensive Organometallic Chemistry III, 2007, Volume 1, Pages 59-92. doi:10.1016/B0-08-045047-4/00002-9
45. R.N. Perutz and T. Braun "Transition Metal-mediated C–F Bond Activation" Comprehensive Organometallic Chemistry III, 2007, Volume 1, Pages 725-758. doi:10.1016/B0-08-045047-4/00028-5.

# Chapter-3
# Fluorine containing Drugs

## 3. Fluorine containing Drugs:

In Below Table give the list of some fluorine containing drugs.

| Sr No. | Active ingredient/ trade name | Structure | Pharmaceutical action | Dosage form |
|---|---|---|---|---|
| 1 | Emtricitabine/Atri pla Emtriva Truvada | 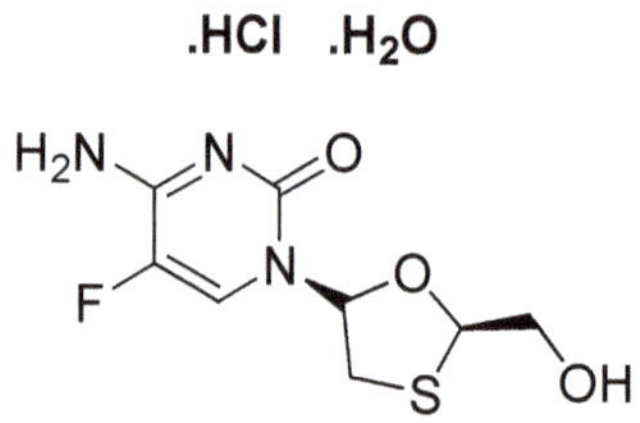 | Antiviral agent Anti - HIV Reverse transcriptase inhibitor | Oral capsule, tablet, solution |
| 2 | Enfl urane/ Entrane | 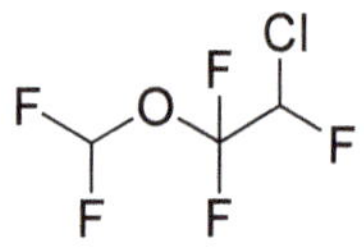 | Anesthetic | Inhalation |
| 3 | Enoxacin Δ / Penetrex | 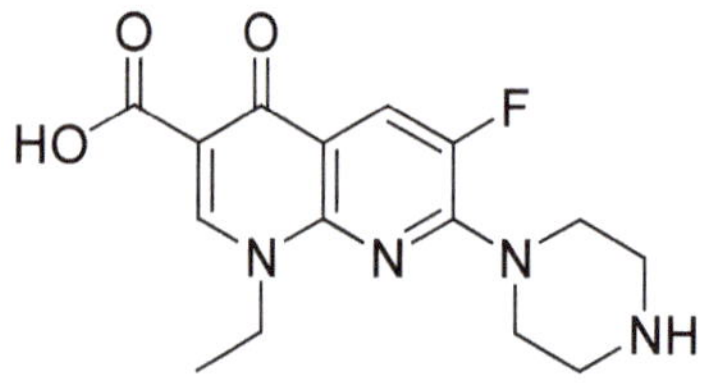 | Anti - infective agent | Oral tablet |
| 4 | Floxuridine/ Fudr | 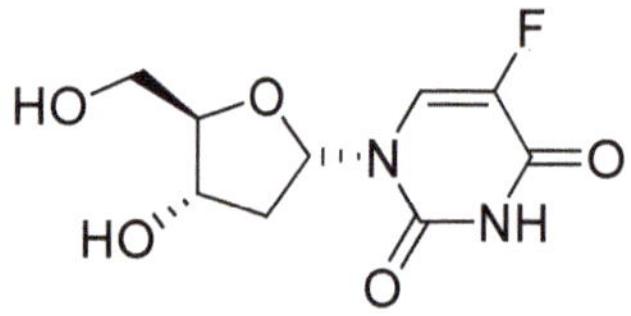 | Antineoplastic, Antiviral | Injectable |
| 5 | Ezetimibe/ Zetia Vytorin | 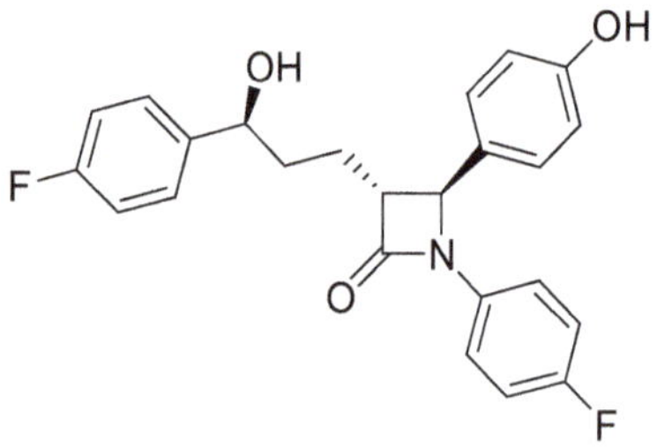 | Antilipemic | Oral tablet |
| 6 | Sevofl uran * / Ultane Sojourn | 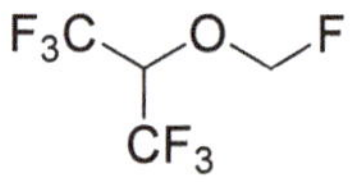 | Platelet aggregation inhibitor Anesthetic | Inhalation |

| 7 | Flucytosine/ Ancobon | 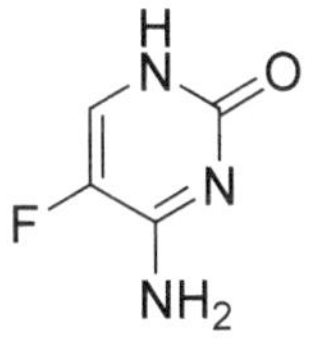 | Antifungal agent | Oral capsule |
|---|---|---|---|---|
| 8 | Fluorouracil/ Fluoroplex Carac Efudex | 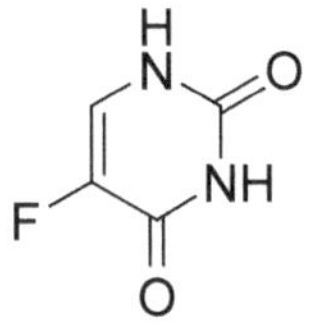 | Antineoplastic Antimetabolite Immunosuppressive agent | Topical solution and cream Injectable |
| 9 | Voriconazole/ Vfend | 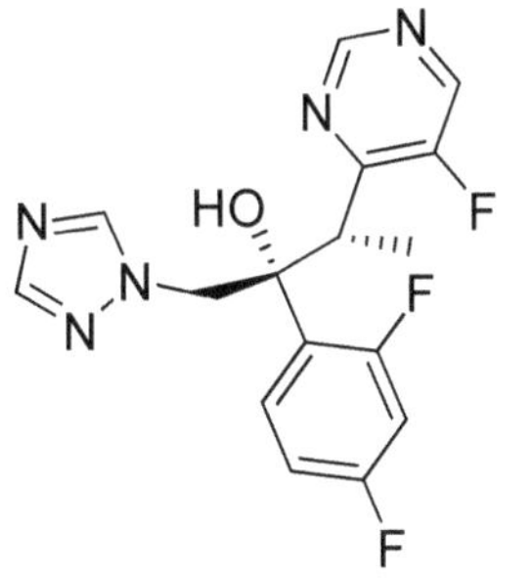 | Antifungal agent | Oral suspension and tablet Injectable |
| 10 | Trifl upromazine 2 / Vesprin Δ | 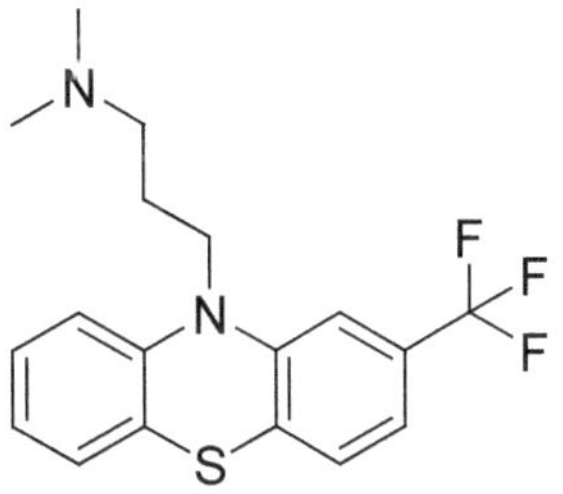 | Antiemetic Antipsychotic agent Dopamine antagonist | Oral suspension and tablet Injectable |
| 11 | Trifl uridine/ Viroptic | 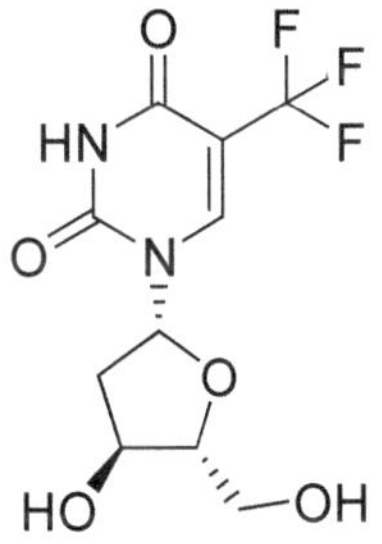 | Antiviral agent Antimetabolite | Ophthalmic drops |
| 12 | Riluzolc/ Rilutck | 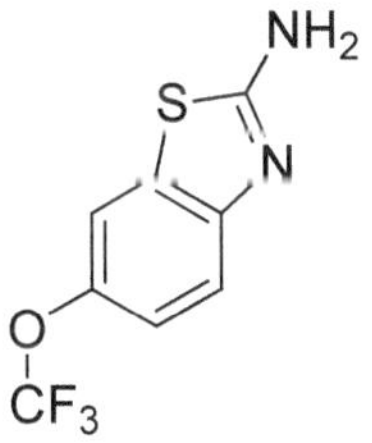 | Anesthetics Anticonvulsant Excitatory amino acid antagonist Neuroprotective agent | Oral tablet |
| 13 | Perfl utren/ Defi nity | 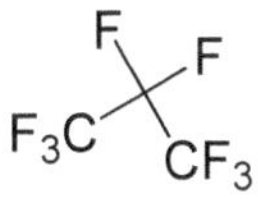 | Contrast media | Injectable |

| | | | | |
|---|---|---|---|---|
| 14 | Trovafl oxacin Mesylate/ Trovan Δ | 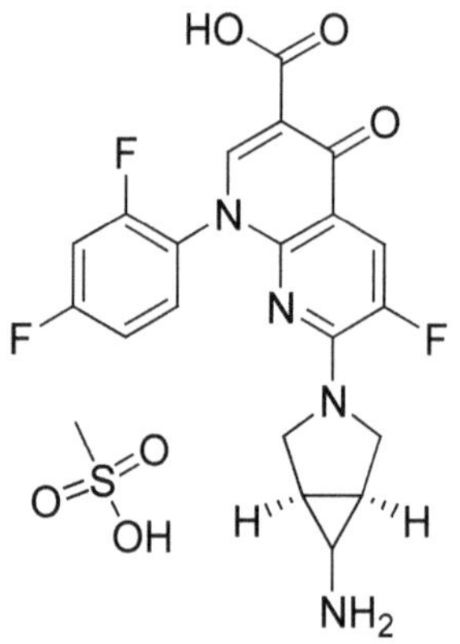 | Anti - infective agent | Oral tablet |
| 15 | Sulindac/ Clinoril | 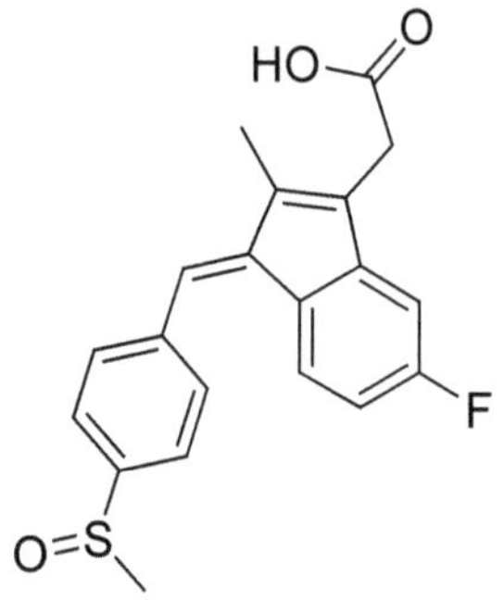 | Nonsteroidal antiinfl ammatory agent Antineoplastic agent Cyclooxygenase inhibitor | Oral tablet |
| 16 | Sparfl oxacin/ Zagam Δ | 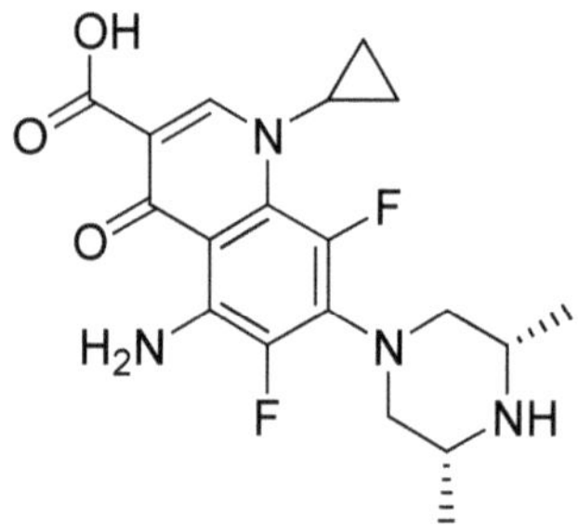 | Antibacterial agent | Oral tablet |
| 17 | Paroxetine Hydrochloride/ Paxil | 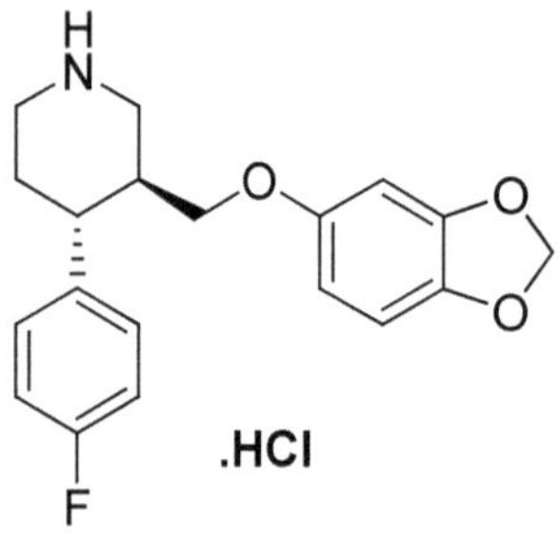 .HCl | Serotonin uptake inhibitor Antidepressive agent Antiobessional agent | Oral tablet and suspension |
| 18 | Paroxetine Mesylate/ Pexeva | 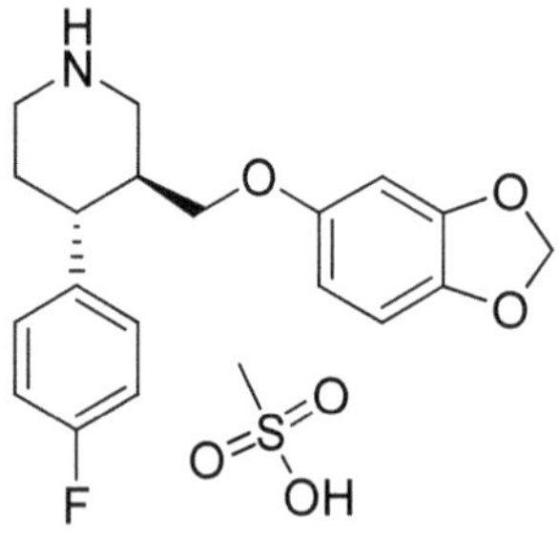 | Antidepressive agent Antiobessional agent | Oral tablet |

| 19 | Quazepam/ Doral | 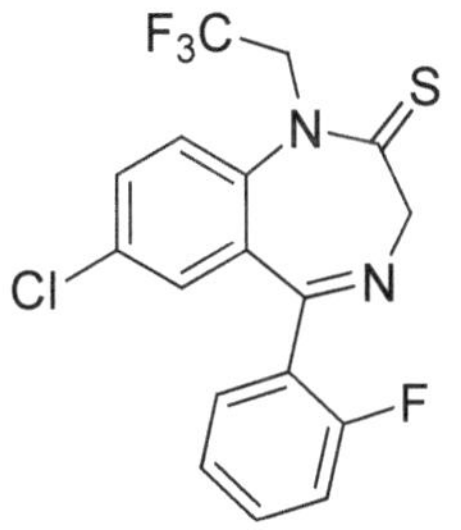 | Hypnotic Sedative | Oral tablet |
|---|---|---|---|---|
| 20 | Perfl exane Δ | 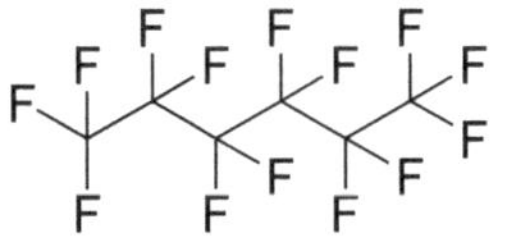 | Preparation of lipid microspheres Imaging agent | Injectable |
| 21 | Perfl ubron/ Imagent Δ | 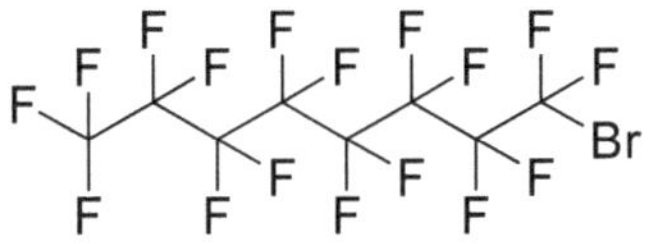 | Contrast media Radiation - sensitizing agent Anti - obesity agent | Oral liquid |
| 22 | Celecoxib/ Celebrex | 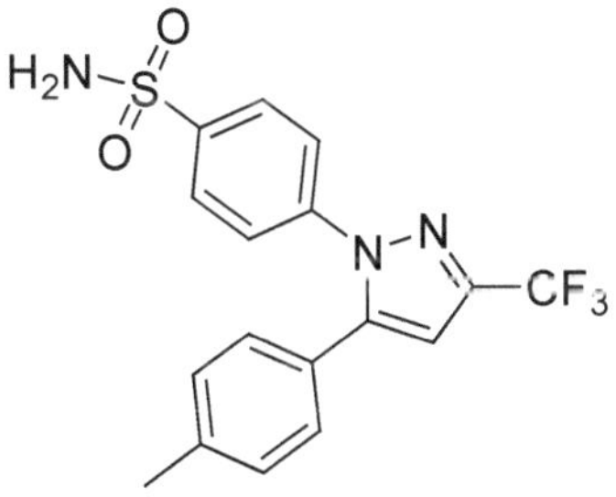 | Anti - infl ammatory Nonsteroidal cyclooxygenase inhibitor | Oral capsule |
| 23 | Cerivastatin Sodium/ Baycol Δ | 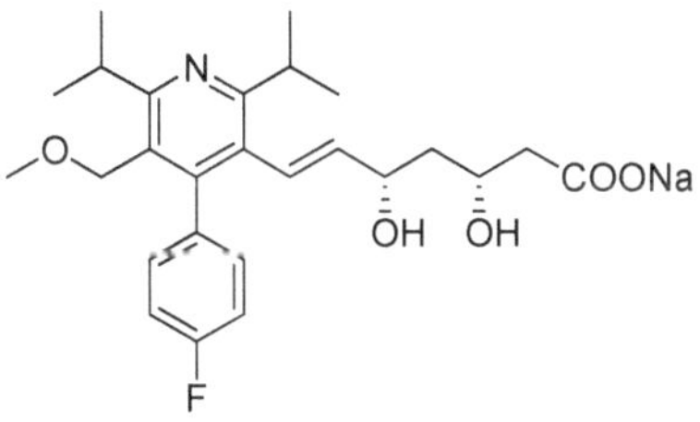 | Hydroxymethylglutaryl -CoA reductase inhibitor | Oral tablet |
| 24 | Cinacalcet Hydrochloride/ Sensipar | 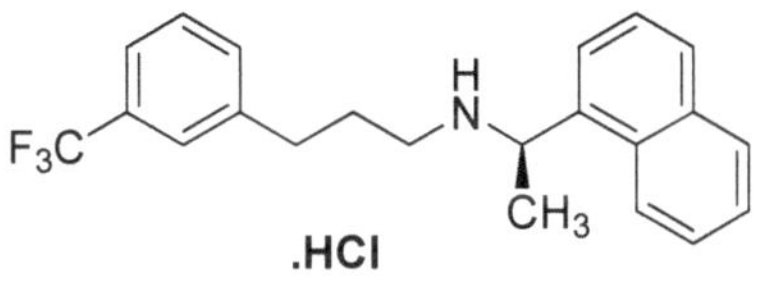 .HCl | Calcimimetic agent | Oral tablet |
| 25 | Clofarabine | 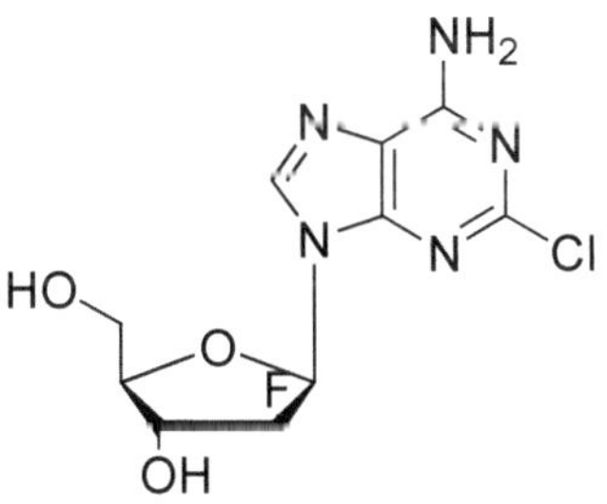 | Anti – metabolite | Injectable |
| 26 | Desfl urane | 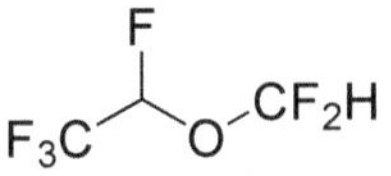 | Anesthetic | Inhalation |

| | | | | |
|---|---|---|---|---|
| 27 | Difl unisal/ Dolobid Δ | 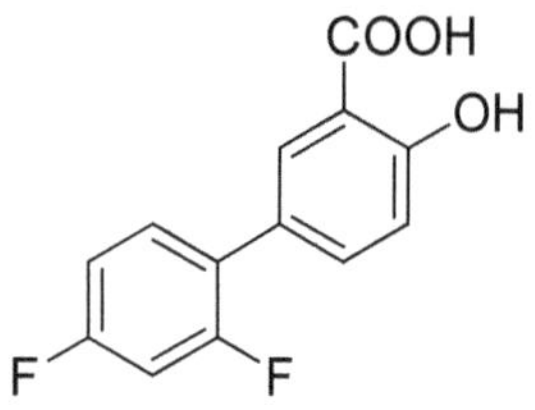 COOH, OH, F, F | Analgesic Anti - Infl ammatory Nonsteroidal cyclooxygenase inhibitor | Oral tablet |
| 28 | Efl ornithine Hydrochloride/ Vaniqa | 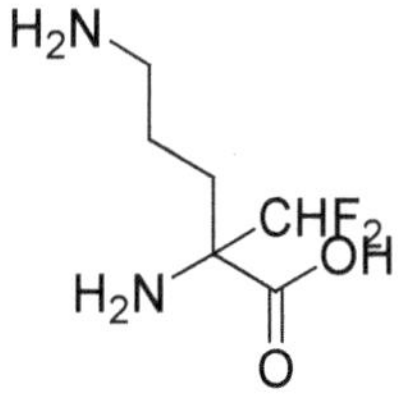 $H_2N$, $CHF_2$, OH, $H_2N$, O | Antineoplastic agent Enzyme inhibitors Trypanocidal agent | Topical cream Injectable Δ |
| 29 | Efavirenz/ Sustiva | 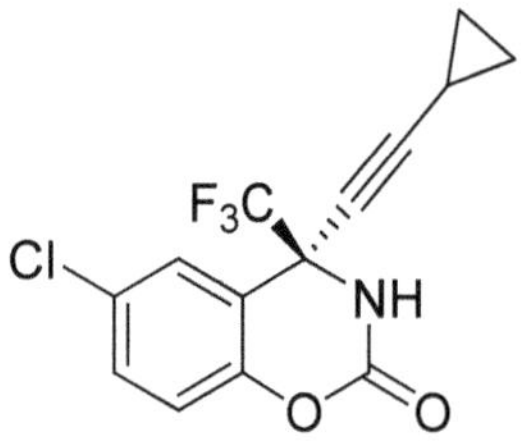 $F_3C$, Cl, NH, O, O | Reverse transcriptase inhibitor Anti - HIV agent | Oral capsule and tablet |
| 30 | Fluconazole/ Difl ucan | 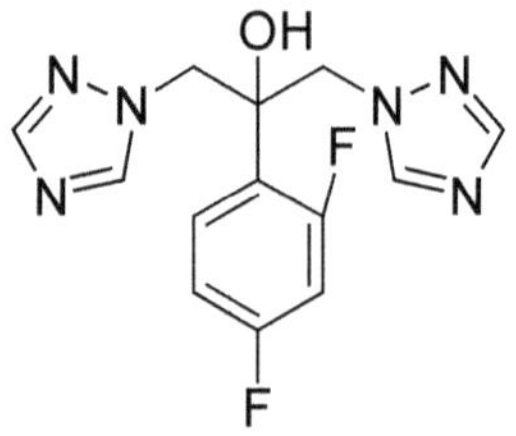 OH, N, N, N, N, N, N, F, F | Antifungal agents | Oral tablet and suspension Injectable |

# Chapter-4
# Water Fluoridation

## 4. Water fluoridation:

### 4.1 History of water fluoridation:

*1909 photograph by Frederick McKay of G.V. Black (left), Isaac Burton and F.Y. Wilson, studying the Colorado brown stain.*[1]

The relationship between fluoride and teeth has been studied since the early 19th century. By 1850, investigators had established that fluoride occurs with varying concentrations in teeth, bone, and drinking water. By 1900, they had speculated that fluoride would protect against tooth decay, proposed supplementing the diet with fluoride, and observed mottled tooth enamel (now called dental fluorosis) without knowing the cause.[2]

The history of water fluoridation can be divided into three periods. The first (c. 1901–1933) was research into the cause of a form of mottled tooth enamel called the Colorado brown stain. The second (c. 1933 1945) focused on the relationship between fluoride concentrations, fluorosis, and tooth decay, and established that moderate levels of fluoride prevent cavities. The third period, from 1945 on, focused on adding fluoride to community water supplies.[3]

The foundation of water fluoridation in the U.S. was the research of the dentist Frederick McKay. McKay spent thirty years investigating the cause of what was then known as the Colorado brown stain, which produced mottled but also cavity-free teeth; with the help of

G.V. Black and other researchers, he established that the cause was fluoride.[4] The first report of a statistical association between the stain and lack of tooth decay was made by UK dentist Norman Ainsworth in 1925. In 1931, an Alcoa chemist, H.V. Churchill, concerned about a possible link between aluminum and staining, analyzed water from several areas where the staining was common and found that fluoride was the common factor.[5]

*H. Trendley Dean set out in 1931 to study fluoride's harm, but by 1950 had demonstrated the cavity-prevention effects of small amounts.[6]*

In the 1930s and early 1940s, H. Trendley Dean and colleagues at the U.S. National Institutes of Health published several epidemiological studies suggesting that a fluoride concentration of about 1 mg/L was associated with substantially fewer cavities in temperate climates, and that it increased fluorosis but only to a level that was of no medical or aesthetic concern. Other studies found no other significant adverse effects even in areas with fluoride levels as high as 8 mg/L.[7] To test the hypothesis that adding fluoride would prevent cavities, Dean and his colleagues conducted a controlled experiment by fluoridating the water in Grand Rapids, Michigan, starting January 25, 1945. The results, published in 1950, showed significant reduction of cavities.[8][9] Significant reductions in tooth decay were also reported by important early studies outside the U.S., including the Brantford–Sarnia–Stratford study in Canada (1945–1962), the Tiel–Culemborg study in the Netherlands (1953–1969), the Hastings study in New Zealand (1954–1970), and the Department of Health study in the U.K. (1955–1960).[5] By present-day standards these and other pioneering studies were crude, but the large reductions in cavities convinced public health professionals of the benefits of fluoridation.[10]

Fluoridation became an official policy of the U.S. Public Health Service by 1951, and by 1960 water fluoridation had become widely used in the U.S., reaching about 50 million people.[7] By 2006, 69.2% of the U.S. population on public water systems were receiving fluoridated water, amounting to 61.5% of the total U.S. population; 3.0% of the population on public water systems were receiving naturally occurring fluoride.[11] In some other countries the pattern was similar. New Zealand, which led the world in per-capita sugar consumption and had the world's worst teeth, began fluoridation in 1953, and by 1968 fluoridation was used by 65% of the population served by a piped water supply.[12] Fluoridation was introduced into Brazil in 1953, was regulated by federal law starting in 1974, and by 2004 was used by 71% of the population.[13] In the Republic of Ireland, fluoridation was legislated in 1960, and after a constitutional challenge the two major cities of Dublin and Cork began it in 1964;[5] fluoridation became required for all sizeable public water systems and by 1996 reached 66% of the population.[14] In other locations, fluoridation was used and then discontinued: in Kuopio, Finland, fluoridation was used for decades but was discontinued because the school dental service provided significant fluoride programs and the cavity risk was low, and in Basel, Switzerland, it was replaced with fluoridated salt.[5]

McKay's work had established that fluorosis occurred before tooth eruption. Dean and his colleagues assumed that fluoride's protection against cavities was also pre-eruptive, and this incorrect assumption was accepted for years. By 2000, however, the topical effects of fluoride (in both water and toothpaste) were well understood, and it had become known that a constant low level of fluoride in the mouth works best to prevent cavities.[10]

This is a very controversial subject. Water fluoridation is a controlled addition of fluoride to public water sources. Many people believe that this is a very good thing for dental health, while many others vehemently disagree by coining it a poisoning of drinking water. Water fluoridation presents an ongoing conflict between “the greater good” and “individual rights”, as it’s addition to water is not consented by everyone who drinks tap water. If fluoride can be harmful to teeth, when its main purpose is to prevent tooth related diseases, one can only imagine what effects ingesting it may have on the human body. There are many researchers, dental and medical professionals, nutritionists and environmentalists that speak out in opposition of water fluoridation.

## 4.2 Recommendations of fluoride:

The Food and Nutrition Board at the Institute of Medicine recommends the following dietary intake for fluoride:

Infants

- 0 - 6 months: 0.01 milligrams per day (mg/day)
- 7 - 12 months: 0.5 mg/day

Children

- 1 - 3 years: 0.7 mg/day
- 4 - 8 years: 1.0 mg/day
- 9 - 13 years: 2.0 mg/day

Adolescents and Adults

- Males ages 14 to 18 years: 3.0 mg/day
- Males over 18 years: 4.0 mg/day
- Females over 14 years: 3.0 mg/day

The best way to get the daily requirement of essential vitamins is to eat a balanced diet that contains a variety of foods from the food guide plate. Specific recommendations depend on age and gender. Ask your health care provider which amount is best for you.

To help make sure infants and children do not get too much fluoride:

- Ask your health care provider about the type of water to use in concentrated or powdered formulas.
- Do not use any fluoride supplement without talking to your health care provider.
- Avoid using fluoride toothpaste in infants younger than 2 years.
- Use only a pea-sized amount of fluoride toothpaste in children older than 2 years.
- Avoid fluoride mouth rinses in children younger than 6 years.

## 4.3 Goal of water fluoridation:

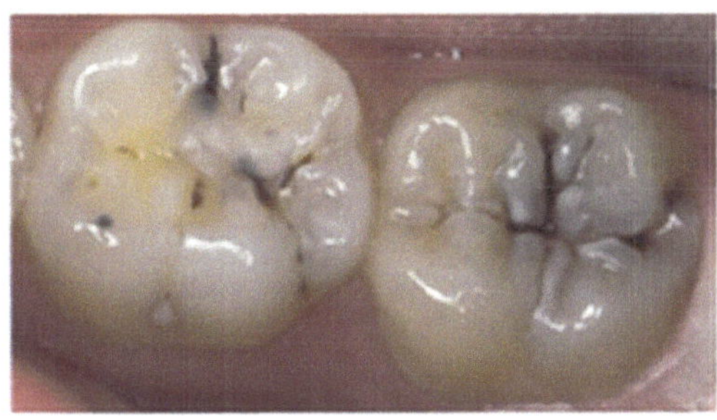

*A cavity starts in a tooth's outer enamel and spreads to the dentin and pulp inside.*

The goal of water fluoridation is to prevent tooth decay by adjusting the concentration of fluoride in public water supplies.[15] Tooth decay (dental caries) is one of the most prevalent chronic diseases worldwide.[16] Although it is rarely life-threatening, tooth decay can cause pain and impair eating, speaking, facial appearance, and acceptance into society,[17] and it greatly affects the quality of life of children, particularly those of low socioeconomic status.[16] In most industrialized countries, tooth decay affects 60–90% of schoolchildren and the vast majority of adults; although the problem appears to be less in Africa's developing countries, it is expected to increase in several countries there because of changing diet and inadequate fluoride exposure.[18] In the U.S., minorities and the poor both have higher rates of decayed and missing teeth,[19] and their children have less dental care.[20] Once a cavity occurs, the tooth's fate is that of repeated restorations, with estimates for the median life of an amalgam tooth filling ranging from 9 to 14 years.[21] Oral disease is the fourth most expensive disease to treat.[22] The motivation for fluoridation of salt or water is similar to that of iodized salt for the prevention of mental retardation and goiter.[23]

The goal of water fluoridation is to prevent a chronic disease whose burdens particularly fall on children and the poor.[16] Its use presents a conflict between the common good and individual rights.[24] It is controversial,[25] and opposition to it has been based on ethical, legal, safety, and efficacy grounds.[26] Health and dental organizations worldwide have endorsed its safety and effectiveness.[27] Its use began in 1945, following studies of children in a region where higher levels of fluoride occur naturally in the water.[28] Researchers discovered that moderate fluoridation prevents tooth decay,[29] and as of 2004 about 400 million people worldwide received fluoridated water.[30]

## 4.4 Water fluoridation Implementation:

*Fluoride monitor (at left) in a community water tower pumphouse, Minnesota, 1987.*

Fluoridation does not affect the appearance, taste, or smell of drinking water.[31] It is normally accomplished by adding one of three compounds to the water: sodium fluoride, fluorosilicic acid, or sodium fluorosilicate

- Sodium fluoride (NaF) was the first compound used and is the reference standard.[32] It is a white, odorless powder or crystal; the crystalline form is preferred if manual handling is used, as it minimizes dust.[33] It is more expensive than the other compounds, but is easily handled and is usually used by smaller utility companies.[34]
- Fluorosilicic acid ($H_2SiF_6$) is the most commonly used additive for water fluoridation.[35] It is an inexpensive liquid by-product of phosphate fertilizer manufacture.[32] It comes in varying strengths, typically 23–25%; because it contains so much water, shipping can be expensive.[33] It is also known as hexafluorosilicic, hexafluosilicic, hydrofluosilicic, and silicofluoric acid.[32]
- Sodium fluorosilicate ($Na_2SiF_6$) is the sodium salt of fluorosilicic acid. It is a powder or very fine crystal that is easier to ship than fluorosilicic acid. It is also known as sodium silicofluoride.[33]

These compounds were chosen for their solubility, safety, availability, and low cost.[32] A 1992 census found that, for U.S. public water supply systems reporting the type of compound used, 63% of the population received water fluoridated with fluorosilicic acid, 28% with sodium fluorosilicate, and 9% with sodium fluoride.[36] The Centers for Disease Control and Prevention developed recommendations for water fluoridation that specify requirements for personnel, reporting, training, inspection, monitoring, surveillance, and actions in case of overfeed, along with technical requirements for each major compound used.[37]

Although fluoride was once considered an essential nutrient, the U.S. National Research Council has since removed this designation due to the lack of studies showing it is essential for human growth, though still considering fluoride a "beneficial element" due to its positive impact on oral health.[38] Since 1962, the U.S. had specified the optimal level of fluoride to range from 0.7 to 1.2 mg/L (milligrams per liter, equivalent to parts per million), depending on the average maximum daily air temperature; the optimal level is lower in warmer climates, where people drink more water, and is higher in cooler climates.[39] This standard, adopted in 1962, is not appropriate for all parts of the world and is based on assumptions that have become obsolete with the rise of air conditioning and increased use of soft drinks, processed food, and other sources of fluorides. In 1994 a World Health Organization expert committee on fluoride use stated that 1.0 mg/L should be an absolute upper bound, even in cold climates, and that 0.5 mg/L may be an appropriate lower limit.[40] A 2007 Australian systematic review recommended a range from 0.6 to 1.1 mg/L.[41] In 2011, the U.S. lowered its recommended level of fluoride to 0.7 mg/L.[42]

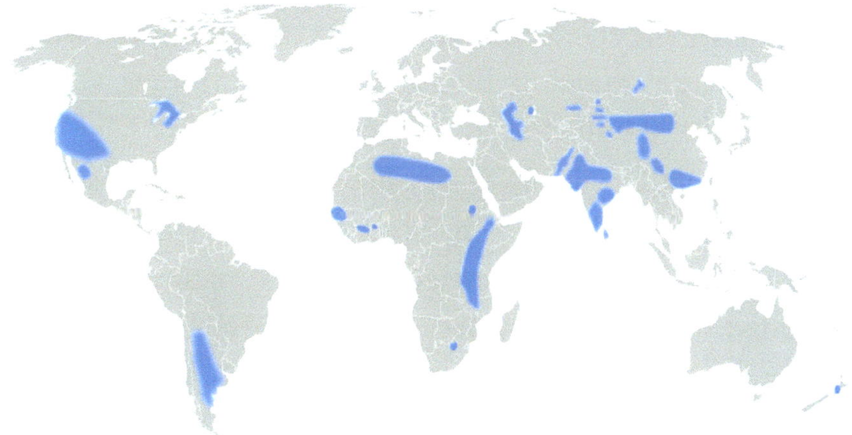

*Geographical areas associated with groundwater having over 1.5 mg/L of naturally occurring fluoride, which is above recommended levels.[41]*

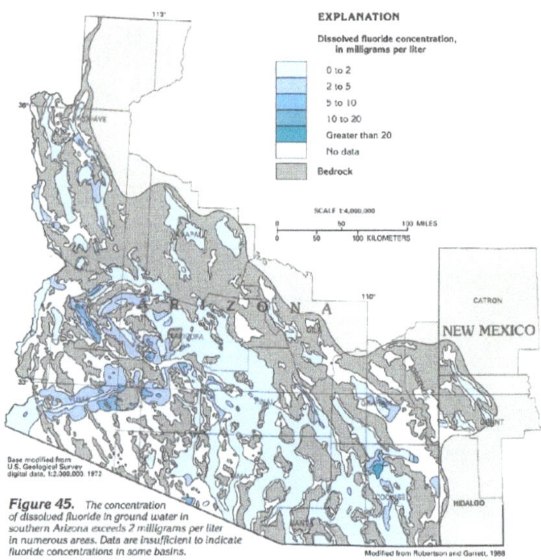


*Detail of southern Arizona. Areas in darker blues have groundwater with over 2 mg/L of naturally occurring fluoride.*

Fluoride naturally occurring in water can be above, at, or below recommended levels. Rivers and lakes generally contain fluoride levels less than 0.5 mg/L, but groundwater, particularly in volcanic or mountainous areas, can contain as much as 50 mg/L.[43] Higher concentrations of fluorine are found in alkaline volcanic, hydrothermal, sedimentary, and other rocks derived from highly evolved magmas and hydrothermal solutions, and this fluorine dissolves into nearby water as fluoride. In most drinking waters, over 95% of total fluoride is the $F^-$ ion, with the magnesium–fluoride complex ($MgF^+$) being the next most common. Because fluoride levels in water are usually controlled by the solubility of fluorite ($CaF_2$), high natural fluoride levels are associated with calcium-deficient, alkaline, and soft waters.[44] Defluoridation is needed when the naturally occurring fluoride level exceeds recommended limits. It can be accomplished by percolating water through granular beds of activated alumina, bone meal, bone char, or tricalcium phosphate; by coagulation with alum; or by precipitation with lime.[45]

Pitcher or faucet-mounted water filters do not alter fluoride; the more-expensive reverse osmosis filters remove 65–95% of fluoride, and distillation filters remove all fluoride.[46] U.S. regulations for bottled water do not require disclosing fluoride content, so the effect of always drinking it is unknown.[46] Surveys of bottled water in Cleveland and in Iowa found that most contained well below optimal fluoride levels;[47] a survey in São Paulo, Brazil, found large variations of fluoride, with many bottles exceeding recommended limits and disagreeing with their labels.[48]

## 4.5 Role of fluoride:

Fluoride exerts its major effect by interfering with the demineralization mechanism of tooth decay. Tooth decay is an infectious disease, the key feature of which is an increase within dental plaque of bacteria such as *Streptococcus mutans* and *Lactobacillus*. These produce organic acids when carbohydrates, especially sugar, are eaten.[49] When enough acid is produced to lower the pH below 5.5,[50] the acid dissolves carbonated hydroxyapatite, the main component of tooth enamel, in a process known as *demineralization*. After the sugar is gone, some of the mineral loss can be recovered—or *remineralized*—from ions dissolved in the saliva. Cavities result when the rate of demineralization exceeds the rate of remineralization, typically in a process that requires many months or years.[49]

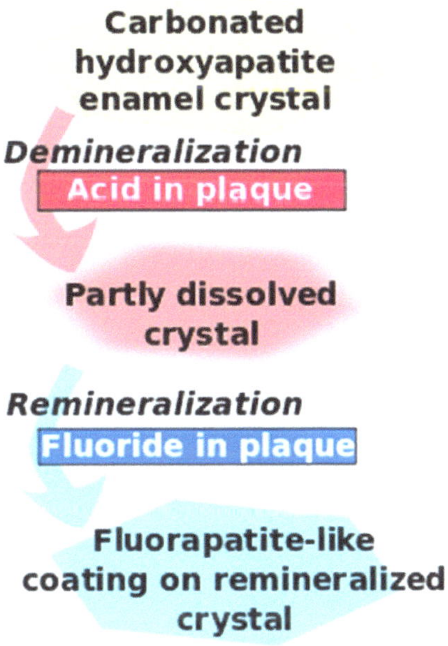

*Demineralization and remineralization of dental enamel in the presence of acid and fluoride in saliva and plaque fluid.[49]*

All fluoridation methods, including water fluoridation, create low levels of fluoride ions in saliva and plaque fluid, thus exerting a topical or surface effect. A person living in an area with fluoridated water may experience rises of fluoride concentration in saliva to about 0.04 mg/L several times during a day.[51] Technically, this fluoride does not prevent cavities but rather controls the rate at which they develop.[52] When fluoride ions are present in plaque fluid along with dissolved hydroxyapatite, and the pH is higher than 4.5,[50] a fluorapatite-like remineralized veneer is formed over the remaining surface of the enamel; this veneer is much more acid-resistant than the original hydroxyapatite, and is formed more quickly than ordinary remineralized enamel would be.[49] The cavity-prevention effect of fluoride is mostly due to these surface effects, which occur during and after tooth eruption.[53] Although some systemic (whole-body) fluoride returns to the saliva via blood plasma, and to unerupted teeth via plasma or crypt fluid, there is little data to determine what percentages of fluoride's anticavity effect comes from these systemic mechanisms.[54] Also, although fluoride affects the physiology of dental bacteria,[55] its effect on bacterial growth does not seem to be relevant to cavity prevention.[56]

Fluoride's effects depend on the total daily intake of fluoride from all sources.[43] About 70–90% of ingested fluoride is absorbed into the blood, where it distributes throughout the body. In infants 80–90% of absorbed fluoride is retained, with the rest excreted, mostly via urine; in adults about 60% is retained. About 99% of retained fluoride is stored in bone, teeth, and other calcium-rich areas, where excess quantities can cause fluorosis.[57] Drinking water is typically the largest source of fluoride.[43] In many industrialized countries swallowed toothpaste is the main source of fluoride exposure in unfluoridated communities.[58] Other sources include dental products other than toothpaste; air pollution from fluoride-containing coal or from phosphate fertilizers; trona, used to tenderize meat in Tanzania; and tea leaves, particularly the tea bricks favored in parts of China. High fluoride levels have been found in other foods, including barley, cassava, corn, rice, taro, yams, and fish protein concentrate. The U.S. Institute of Medicine has established Dietary Reference Intakes for fluoride: Adequate Intake values range from 0.01 mg/day for infants aged 6 months or less, to 4 mg/day for men aged 19 years and up; and the Tolerable Upper Intake Level is 0.10 mg/kg/day for infants and children through age 8 years, and 10 mg/day thereafter.[59] A rough estimate is that an adult in a temperate climate consumes 0.6 mg/day of fluoride without fluoridation, and 2 mg/day with fluoridation. However, these values differ greatly among the world's regions: for example, in Sichuan, China the average daily fluoride intake is only 0.1 mg/day in drinking water but 8.9 mg/day in food and 0.7 mg/day directly from the air due to the use of high-fluoride soft coal for cooking and drying foodstuffs indoors.[43]

### 4.6 Effect of water fluoridation:

Water fluoridation effectively reduces cavities in both children and adults:[60] earlier studies showed that water fluoridation reduced childhood cavities by fifty to sixty percent, but more recent studies show lower reductions (18–40%) likely due to increasing use of fluoride from other sources, notably toothpaste, and also the 'halo effect' of food and drink that is made in fluoridated areas and consumed in unfluoridated ones.[15]

A 2000 systematic review found that water fluoridation was statistically associated with a decreased proportion of children with cavities (the median of mean decreases was 14.6%, the range −5 to 64%), and with a decrease in decayed, missing, and filled primary teeth (the median of mean decreases was 2.25 teeth, the range 0.5–4.4 teeth),[61] which is roughly equivalent to preventing 40% of cavities.[62] The review found that the evidence was of moderate quality: few studies attempted to reduce observer bias, control for confounding factors, report variance measures, or use appropriate analysis. Although no major differences between natural and artificial fluoridation were apparent, the evidence was inadequate for a conclusion about any differences.[61] Fluoride also prevents cavities in adults of all ages. There are fewer studies in adults however, and the design of water fluoridation studies in adults is inferior to that of studies of self- or clinically applied fluoride. A 2007 meta-analysis found that water fluoridation prevented an estimated 27% of cavities in adults (95% confidence interval [CI] 19–34%), about the same fraction as prevented by exposure to any delivery method of fluoride (29% average, 95% CI: 16–42%).[63] A 2002 systematic review found strong evidence that water fluoridation is effective at reducing overall tooth decay in communities.[64]

Most countries in Europe have experienced substantial declines in cavities without the use of water fluoridation.[51] For example, in Finland and Germany, tooth decay rates remained stable or continued to decline after water fluoridation stopped. Fluoridation may be useful in the U.S. because unlike most European countries, the U.S. does not have school-based dental care, many children do not visit a dentist regularly, and for many U.S. children water fluoridation is the prime source of exposure to fluoride.[65] The effectiveness of water fluoridation can vary according to circumstances such as whether preventive dental care is free to all children.[66]

Some studies suggest that fluoridation reduces oral health inequalities between the rich and poor, but the evidence is limited.[51] There is anecdotal but not scientific evidence that fluoride allows more time for dental treatment by slowing the progression of tooth decay, and that it simplifies treatment by causing most cavities to occur in pits and fissures of teeth.[67]

### 4.7 Adverse effect of fluoride:

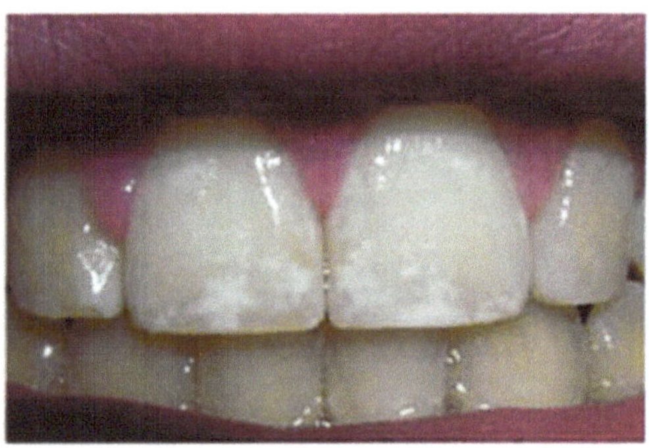

*A mild case of dental fluorosis, visible as white streaks on the subject's upper right central incisor.*

Fluoride's adverse effects depend on total fluoride dosage from all sources. At the commonly recommended dosage, the only clear adverse effect is dental fluorosis, which can alter the appearance of children's teeth during tooth development; this is mostly mild and is unlikely to represent any real effect on aesthetic appearance or on public health.[41] The critical period of exposure is between ages one and four years, with the risk ending around age eight. Fluorosis can be prevented by monitoring all sources of fluoride, with fluoridated water directly or indirectly responsible for an estimated 40% of risk and other sources, notably toothpaste, responsible for the remaining 60%.[68] Compared to water naturally fluoridated at 0.4 mg/L, fluoridation to 1 mg/L is estimated to cause additional fluorosis in one of every 6 people (95% CI 4–21 people), and to cause additional fluorosis of aesthetic concern in one of every 22 people (95% CI 13.6–∞ people). Here, *aesthetic concern* is a term used in a standardized scale based on what adolescents would find unacceptable, as measured by a 1996 study of British 14-year-olds.[61] In many industrialized countries the prevalence of fluorosis is increasing even in unfluoridated communities, mostly because of fluoride from swallowed toothpaste.[58] A 2009 systematic review indicated that fluorosis is associated with consumption of infant formula or of water added to reconstitute the formula, that the evidence was distorted by publication bias, and that the evidence that the formula's fluoride caused the fluorosis was weak.[69] In the U.S. the decline in tooth decay was accompanied by increased fluorosis in both fluoridated and unfluoridated communities; accordingly, fluoride has been reduced in various ways worldwide in infant formulas, children's toothpaste, water, and fluoride-supplement schedules.[67]

Fluoridation has little effect on risk of bone fracture (broken bones); it may result in slightly lower fracture risk than either excessively high levels of fluoridation or no fluoridation.[41] There is no clear association between fluoridation and cancer or deaths due to cancer, both for cancer in general and also specifically for bone cancer and osteosarcoma.[41][61] Other adverse effects lack sufficient evidence to reach a confident conclusion.[61] A Finnish study published in 1997 showed that fear that water is fluoridated may have a psychological effect with a large variety of symptoms, regardless of whether the water is actually fluoridated.[31]

Fluoride can occur naturally in water in concentrations well above recommended levels, which can have several long-term adverse effects, including severe dental fluorosis, skeletal fluorosis, and weakened bones.[57] The World Health Organization recommends a guideline maximum fluoride value of 1.5 mg/L as a level at which fluorosis should be minimal.[70] In rare cases improper implementation of water fluoridation can result in overfluoridation that causes outbreaks of acute fluoride poisoning, with symptoms that include nausea, vomiting, and diarrhea. Three such outbreaks were reported in the U.S. between 1991 and 1998, caused by fluoride concentrations as high as 220 mg/L; in the 1992 Alaska outbreak, 262 people became ill and one person died.[71] In 2010, approximately 60 gallons of fluoride were released into the water supply in Asheboro, North Carolina in 90 minutes—an amount that was intended to be released in a 24-hour period.[72]

Like other common water additives such as chlorine, hydrofluosilicic acid and sodium silicofluoride decrease pH and cause a small increase of corrosivity, but this problem is easily addressed by increasing the pH.[73] Although it has been hypothesized that hydrofluosilicic acid and sodium silicofluoride might increase human lead uptake from water, a 2006 statistical analysis did not support concerns that these chemicals cause higher blood lead concentrations in children.[74] Trace levels of arsenic and lead may be present in fluoride compounds added to water, but no credible evidence exists that their presence is of concern: concentrations are below measurement limits.[73]

The effect of water fluoridation on the natural environment has been investigated, and no adverse effects have been established. Issues studied have included fluoride concentrations in groundwater and downstream rivers; lawns, gardens, and plants; consumption of plants grown in fluoridated water; air emissions; and equipment noise.[73]

### 4.8 Alternatives of water fluoridation:

Although water fluoridation is the most effective means of achieving fluoride exposure that is community-wide,[41] other fluoride therapies are also effective in preventing tooth decay;[16] they include fluoride toothpaste, mouthwash, gel, and varnish,[75] and fluoridation of salt and milk.[76] Dental sealants are effective as well,[16] with estimates of prevented cavities ranging from 33% to 86%, depending on age of sealant and type of study.[75]

*Fluoride toothpaste is effective against cavities. It is widely used, but less so among the poor.[76]*

Fluoride toothpaste is the most widely used and rigorously evaluated fluoride treatment.[76] Its introduction in the early 1970s is considered the main reason for the decline in tooth decay in industrialized countries,[51] and toothpaste appears to be the single common factor in countries where tooth decay has declined.[77] Toothpaste is the only realistic fluoride strategy in many low-income countries, where lack of infrastructure renders water or salt fluoridation infeasible.[78] However, it relies on individual and family behavior, and its use is less likely among lower economic classes;[76] in low-income countries it is unaffordable for the poor.[78] Fluoride toothpaste prevents about 25% of cavities in young permanent teeth, and its effectiveness is improved if higher concentrations of fluoride are used, or if the toothbrushing is supervised. Fluoride mouthwash and gel are about as effective as fluoride toothpaste; fluoride varnish prevents about 45% of cavities.[75] By comparison, brushing with nonfluoride toothpaste has little effect on cavities.[58]

The effectiveness of salt fluoridation is about the same as that of water fluoridation, if most salt for human consumption is fluoridated. Fluoridated salt reaches the consumer in salt at home, in meals at school and at large kitchens, and in bread. For example, Jamaica has just one salt producer, but a complex public water supply; it started fluoridating all salt in 1987, achieving a notable decline in cavities. Universal salt fluoridation is also practiced in Colombia and the Swiss Canton of Vaud; in Germany fluoridated salt is widely used in households but unfluoridated salt is also available, giving consumers choice about fluoride. Concentrations of fluoride in salt range from 90 to 350 mg/kg, with studies suggesting an optimal concentration of around 250 mg/kg.[76]

Milk fluoridation is practiced by the Borrow Foundation in some parts of Bulgaria, Chile, Peru, Russia, Macedonia, Thailand and the UK. Depending on location, the fluoride is added to milk, to powdered milk, or to yogurt. For example, milk-powder fluoridation is used in rural Chilean areas where water fluoridation is not technically feasible.[79] These programs are aimed at children, and have neither targeted nor been evaluated for adults.[76] A 2005 systematic review found insufficient evidence to support the practice, but also concluded that studies suggest that fluoridated milk benefits schoolchildren, especially their permanent teeth.[80]

Other public-health strategies to control tooth decay, such as education to change behavior and diet, have lacked impressive results.[67] Although fluoride is the only well-documented agent which controls the rate at which cavities develop, it has been suggested that adding calcium to the water would reduce cavities further.[81] Other agents to prevent tooth decay include antibacterials such as chlorhexidine and sugar substitutes such as xylitol.[75] Xylitol-sweetened chewing gum has been recommended as a supplement to fluoride and other conventional treatments if the gum is not too costly.[82] Two proposed approaches, bacteria replacement therapy (probiotics) and caries vaccine, would share water fluoridation's advantage of requiring only minimal patient compliance, but have not been proven safe and effective.[75] Other experimental approaches include fluoridated sugar, polyphenols, and casein phosphopeptide–amorphous calcium phosphate nanocomplexes.[83]

A 2007 Australian review concluded that water fluoridation is the most effective and socially the most equitable way to expose entire communities to fluoride's cavity-prevention effects.[41] A 2002 U.S. review estimated that sealants decreased cavities by about 60% overall, compared to about 18–50% for fluoride.[64] A 2007 Italian review suggested that water fluoridation may not be needed, particularly in the industrialized countries where cavities have become rare, and concluded that toothpaste and other topical fluoride offers a best way to prevent cavities worldwide.[51] A 2004 World Health Organization review stated that water fluoridation, when it is culturally acceptable and technically feasible, has substantial advantages in preventing tooth decay, especially for subgroups at high risk.[18]

### 4.9 Water fluoridation expenditure:

Fluoridation costs an estimated $1.02 per person-year on the average (range: $0.24 $10.79; all costs in this paragraph are for the U.S.[15] and are in 2014 dollars, inflation-adjusted from earlier estimates[84]). Larger water systems have lower per capita cost, and the cost is also affected by the number of fluoride injection points in the water system, the type of feeder and monitoring equipment, the fluoride chemical and its transportation and storage, and water plant personnel expertise.[15] In affluent countries the cost of salt fluoridation is also negligible; developing countries may find it prohibitively expensive to import the fluoride additive.[85] By comparison, fluoride toothpaste costs an estimated \$8–\$17 per person-year, with the incremental cost being zero for people who already brush their teeth for other reasons; and dental cleaning and application of fluoride varnish or gel costs an estimated \$93 per person-year. Assuming the worst case, with the lowest estimated effectiveness and highest estimated operating costs for small cities, fluoridation costs an estimated \$16–\$24 per saved tooth-decay surface, which is lower than the estimated \$92 to restore the surface[15] and the estimated \$156 average discounted lifetime cost of the decayed surface, which includes the cost to maintain the restored tooth surface.[21] It is not known how much is spent in industrial countries to treat dental fluorosis, which is mostly due to fluoride from swallowed toothpaste.[58]

Although a 1989 workshop on cost-effectiveness of cavity prevention concluded that water fluoridation is one of the few public health measures that save more money than they cost, little high-quality research has been done on the cost-effectiveness and solid data are scarce.[15][39] Dental sealants are cost-effective only when applied to high-risk children and teeth.[32] A 2002 U.S. review estimated that on average, sealing first permanent molars saves costs when they are decaying faster than 0.47 surfaces per person-year whereas water fluoridation saves costs when total decay incidence exceeds 0.06 surfaces per person-year.[64] In the U.S., water fluoridation is more cost-effective than other methods to reduce tooth decay in children, and a 2008 review concluded that water fluoridation is the best tool for combating cavities in many countries, particularly among socially disadvantaged groups.[67]

U.S. data from 1974 to 1992 indicate that when water fluoridation is introduced into a community, there are significant decreases in the number of employees per dental firm and the number of dental firms. The data suggest that some dentists respond to the demand shock by moving to non-fluoridated areas and by retraining as specialists.[86]

### 4.10 Principles and policy of water fluoridation:

Like vaccination and food fortification, fluoridation pits the common good against individual rights.[24] Fluoridation can be viewed as a violation of ethical or legal rules that prohibit medical treatment without medical supervision or informed consent, and that prohibit administration of unlicensed medical substances.[51] It can also be viewed as a public health intervention, replicating the benefits of naturally fluoridated water, which can free people from the misery and expense of tooth decay and toothache, with the greatest benefit accruing to those least able to help themselves. This perspective suggests it would be unethical to withhold such treatment.[87]

National and international health agencies and dental associations throughout the world have endorsed water fluoridation as safe and effective.[51][88] The Centers for Disease Control and Prevention listed water fluoridation as one of the ten great public health achievements of the 20th century,[89] along with vaccination, family planning, recognition of the dangers of smoking, and other achievements.[90] Other organizations endorsing fluoridation include the World Health Organization,[18][22] the U.S. Surgeon General,[91] the American Public Health Association,[92] the European Academy of Paediatric Dentistry,[93] and the national dental associations of Australia,[94] Canada,[95] and the U.S.[96]

Despite support by public health organizations and authorities, efforts to introduce water fluoridation have met considerable opposition. Antifluoridation arguments are, "often based on Internet resources or books that present a highly misleading picture of water fluoridation".[97] Fluoridation began during a time of great optimism and faith in science and experts (the 1950s and 1960s), but even then, the public frequently objected. Opponents drew on distrust of experts and unease about medicine and science.[98] Controversies include disputes over fluoridation's benefits and the strength of the evidence basis for these benefits, the difficulty of identifying harms, legal issues over whether water fluoride is a medicine, and the ethics of mass intervention.[99] U.S. opponents of fluoridation were heartened by a 2006 National Research Council report about hazards of water naturally fluoridated to high levels;[100] the report recommended lowering the U.S. maximum limit of 4 mg/L for fluoride in drinking water.[101] Opposition campaigns involve newspaper articles, talk radio, and public forums. Media reporters are often poorly equipped to explain the scientific issues, and are motivated to present controversy regardless of the underlying scientific merits. Websites, which are increasingly used by the public for health information, contain a wide range of material about fluoridation ranging from factual to fraudulent, with a disproportionate percentage opposed to fluoridation. Antifluoridationist literature links fluoride exposure to a wide variety of effects, including AIDS, allergy, Alzheimer's disease, arthritis, cancer, and low IQ, along with diseases of the gastrointestinal tract, kidney, pineal gland, and thyroid.[97]

## At the Sign of THE UNHOLY THREE

*Illustration in a 1955 flier by the Keep America Committee, alleging that fluoridation was a Communist plot.*

Opponents of fluoridation include some researchers, dental and medical professionals, alternative medical practitioners such as chiropractors, health food enthusiasts, a few religious groups (mostly Christian Scientists in the U.S.), and occasionally consumer groups and environmentalists.[102] Organized political opposition has come from libertarians,[103] the John Birch Society,[104] and from groups like the Green parties in the UK and New Zealand.[105][106] Many people do not know that fluoridation is meant to prevent tooth decay, or that natural or bottled water can contain fluoride. As fluoridation does not appear to be an important issue for the general public in the U.S., the debate may reflect an argument between two relatively small lobbies for and against fluoridation.[107] A 2009 survey of Australians found that 70% supported and 15% opposed fluoridation. Those opposed were much more likely to score higher on outrage factors such as "unclear benefits".[108] A 2003 study of focus groups from 16 European countries found that fluoridation was opposed by a majority of focus group members in most of the countries, including France, Germany, and the UK.[107] A 1999 survey in Sheffield, UK found that while a 62% majority favored water fluoridation in the city, the 31% that were opposed expressed their preference with greater intensity than supporters.[109] A 2007 Scottish bioethics council report concluded that good evidence for or against water fluoridation is lacking, therefore local and regional democratic procedures are the most appropriate way to decide whether to fluoridate.[110] Every year in the U.S., pro- and anti-fluoridationists face off in referenda or other public decision-making processes: in most of them, fluoridation is rejected.[102] In the U.S., rejection is more likely when the decision is made by a public referendum; in Europe, most decisions against fluoridation have been made administratively.[111] Neither side of the dispute appears to be weakening or willing to concede.[102]

Conspiracy theories involving fluoridation are common, and include claims that fluoridation was motivated by protecting the U.S. atomic bomb program from litigation, that (as famously parodied in the film *Dr. Strangelove,* where a deranged U.S. Army general claimed that it would "sap and impurify all of our precious bodily fluids") it is part of a Communist or New World Order plot to take over the world, that it was pioneered by a German chemical company to make people submissive to those in power, that behind the scenes it is promoted by the sugary food or phosphate fertilizer or aluminum industries, or that it is a smokescreen to cover failure to provide dental care to the poor.[97] One such theory is that fluoridation was a public-relations ruse sponsored by fluoride polluters such as the aluminum maker Alcoa and the Manhattan Project, with conspirators that included industrialist Andrew Mellon and the Mellon Institute's researcher Gerald J. Cox, the Kettering Laboratory of the University of Cincinnati, the Federal Security Agency's administrator Oscar R. Ewing, and public-relations strategist Edward Bernays.[112] Specific antifluoridation arguments change to match the spirit of the time.[113]

## 4.11 Reference:


1. Douglas WA. *History of Dentistry in Colorado, 1859–1959.* Denver: Colorado State Dental Assn; 1959. OCLC 5015927. p. 199.
2. Cox GJ. Fluorine and dental caries. In: Toverud G, Finn SB, Cox GJ, Bodecker CF, Shaw JH, editors. *A Survey of the Literature of Dental Caries.* Washington, DC: National Academy of Sciences—National Research Council; 1952. Publication 225. OCLC 14681626. p. 325–414.
3. Ripa LW. A half-century of community water fluoridation in the United States: review and commentary [PDF]. *J Public Health Dent.* 1993; 53(1):17–44. doi:10.1111/j.1752-7325.1993.tb02666.x. PMID 8474047.
4. Colorado brown stain:
    i. Peterson J. Solving the mystery of the Colorado Brown Stain. *J Hist Dent.* 1997; 45(2):57–61. PMID 9468893.
    ii. Colorado Springs Dental Society. The discovery of fluoride; 2004 [Retrieved 2012-06-11].
5. Mullen J. History of water fluoridation. *Br Dent J.* 2005; 199(7s):1–4. doi:10.1038/sj.bdj.4812863. PMID 16215546.
6. Division of Oral Health, National Center for Chronic Disease Prevention and Health Promotion, CDC. Achievements in public health, 1900–1999: Fluoridation

of drinking water to prevent dental caries. *MMWR Morb Mortal Wkly Rep.* 1999; 48(41):933–40. Contains H. Trendley Dean, D.D.S. Reprinted in: *JAMA.* 2000; 283(10):1283–6. doi:10.1001/jama. 283.10.1283. PMID 10714718.

7. Lennon MA. One in a million: the first community trial of water fluoridation. *Bull World Health Organ.* 2006; 84(9):759–60. doi:10.2471/ BLT.05.028209. PMID 17128347. PMC 2627472.
8. National Institute of Dental and Craniofacial Research. The story of fluoridation; 2008-12-20 [Retrieved 2010-02-06].
9. Dean HT, Arnold FA, Jay P, Knutson JW. Studies on mass control of dental caries through fluoridation of the public water supply. *Public Health Rep.* 1950; 65(43):1403–8. PMID 14781280.
10. Burt BA, Tomar SL. Changing the face of America: water fluoridation and oral health. In: Ward JW, Warren C. *Silent Victories: The History and Practice of Public Health in Twentieth-century America.* Oxford University Press; 2007. ISBN 0-19-515069-4. p. 307–22.
11. Division of Oral Health, National Center for Chronic Disease Prevention and Health Promotion, CDC. Water fluoridation statistics for 2006; 2008-09-17 [Retrieved 2008-12-22].
12. Akers HF. Collaboration, vision and reality: water fluoridation in New Zealand (1952–1968) [PDF]. *N Z Dent J.* 2008; 104(4):127–33. PMID 19180863.
13. Buzalaf MA, de Almeida BS, Olympio KPK, da S Cardoso VE, de CS Peres SH. Enamel fluorosis prevalence after a 7-year interruption in water fluoridation in Jaú, São Paulo, Brazil. *J Public Health Dent.* 2004; 64(4):205–8. doi:10.1111/j.1752-7325.2004.tb02754.x. PMID 15562942.
14. The British Fluoridation Society; The UK Public Health Association; The British Dental Association; The Faculty of Public Health. *One in a Million: The facts about water fluoridation.* 2nd ed. Manchester: British Fluoridation Society; 2004. ISBN 0-9547684-0-X. The extent of water fluoridation [PDF]. p. 55–80.
15. Centers for Disease Control and Prevention. Recommendations for using fluoride to prevent and control dental caries in the United States. *MMWR Recomm Rep.* 2001; 50(RR-14):1–42. PMID 11521913. Lay summary: *CDC*, 2007-08-09.
16. Selwitz RH, Ismail AI, Pitts NB. Dental caries. *Lancet.* 2007; 369(9555):51–9. doi:10.1016/S0140-6736(07)60031-2. PMID 17208642.
17. Gibson-Moore H. Water fluoridation for some—should it be for all?. *Nutr Bull.* 2009; 34(3):291–5. doi:10.1111/j.1467-3010.2009.01762.x.

18. Petersen PE, Lennon MA. Effective use of fluorides for the prevention of dental caries in the 21st century: the WHO approach [PDF]. *Community Dent Oral Epidemiol*. 2004; 32(5):319–21. doi:10.1111/j.1600-0528.2004. 00175.x. PMID 15341615.
19. Hudson K, Stockard J, Ramberg Z. The impact of socioeconomic status and race-ethnicity on dental health. *Sociol Perspect*. 2007; 50(1):7–25. doi:10.1525/sop. 2007.50.1.7.
20. Vargas CM, Ronzio CR. Disparities in early childhood caries. *BMC Oral Health*. 2006; 6(Suppl 1):S3. doi:10.1186/1472-6831-6-S1-S3. PMID 16934120. PMC 2147596.
21. Griffin SO, Jones K, Tomar SL. An economic evaluation of community water fluoridation [PDF]. *J Public Health Dent*. 2001; 61(2):78–86. doi:10.1111/j.1752-7325.2001.tb03370.x. PMID 11474918.
22. Petersen PE. World Health Organization global policy for improvement of oral health—World Health Assembly 2007 [PDF]. *Int Dent J*. 2008; 58(3):115–21. PMID 18630105.
23. Horowitz HS. Decision-making for national programs of community fluoride use. *Community Dent Oral Epidemiol*. 2000; 28(5):321–9. doi:10.1034/j.1600-0528.2000.028005321.x. PMID 11014508.
24. Ethics:
    i. McNally M, Downie J. The ethics of water fluoridation. *J Can Dent Assoc*. 2000; 66(11):592–3. PMID 11253350.
    ii. Cohen H, Locker D. The science and ethics of water fluoridation. *J Can Dent Assoc*. 2001; 67(10):578–80. PMID 11737979.
25. Cheng KK, Chalmers I, Sheldon TA. Adding fluoride to water supplies [PDF]. *BMJ*. 2007; 335(7622):699–702. doi:10.1136/bmj.39318.562951. BE. PMID 17916854. PMC 2001050.
26. Armfield JM. When public action undermines public health: a critical examination of antifluoridationist literature. *Aust New Zealand Health Policy*. 2007; 4:25. doi:10.1186/1743-8462-4-25. PMID 18067684. PMC 2222595.
27. Pizzo G, Piscopo MR, Pizzo I, Giuliana G. Community water fluoridation and caries prevention: a critical review. *Clin Oral Investig*. 2007; 11(3):189–93. doi:10.1007/s00784-007-0111-6. PMID 17333303.
28. National Institute of Dental and Craniofacial Research. The story of fluoridation; 2008-12-20 [Retrieved 2010-02-06].

29. Ripa LW. A half-century of community water fluoridation in the United States: review and commentary [PDF]. *J Public Health Dent.* 1993; 53(1):17–44. doi:10.1111/j.1752-7325.1993.tb02666.x. PMID 8474047.
30. The British Fluoridation Society; The UK Public Health Association; The British Dental Association; The Faculty of Public Health. *One in a Million: The facts about water fluoridation.* 2nd ed. Manchester: British Fluoridation Society; 2004. ISBN 0-9547684-0-X. The extent of water fluoridation [PDF]. p. 55–80.
31. Lamberg M, Hausen H, Vartiainen T. Symptoms experienced during periods of actual and supposed water fluoridation. *Community Dent Oral Epidemiol.* 1997; 25(4):291–5. doi:10.1111/j.1600-0528.1997.tb00942.x. PMID 9332806.
32. Reeves TG. Centers for Disease Control. Water fluoridation: a manual for engineers and technicians [PDF]; 1986 [Retrieved 2008-12-10].
33. Lauer WC. *Water Fluoridation Principles and Practices.* 5th ed. Vol. M4. American Water Works Association; 2004. (Manual of Water Supply Practices). ISBN 1-58321-311-2. History, theory, and chemicals. p. 1–14.
34. Nicholson JW, Czarnecka B. Fluoride in dentistry and dental restoratives. In: Tressaud A, Haufe G, editors. *Fluorine and Health.* Elsevier; 2008. ISBN 978-0-444-53086-8. p. 333–78.
35. http://www.cdc.gov/fluoridation/faqs/additives.htm#a1
36. Division of Oral Health, National Center for Prevention Services, CDC. Fluoridation census 1992 [PDF]. 1993 [Retrieved 2008-12-29].
37. Centers for Disease Control and Prevention. Engineering and administrative recommendations for water fluoridation, 1995. *MMWR Recomm Rep.* 1995;44(RR-13):1–40. PMID 7565542.
38. Burt BA (May 1992). "The changing patterns of systemic fluoride intake". *J. Dent. Res.* **71** (5): 1228–37. doi:10.1177/00220345920710051601. PMID 1607439.
39. Bailey W, Barker L, Duchon K, Maas W. Populations receiving optimally fluoridated public drinking water—United States, 1992–2006. *MMWR Morb Mortal Wkly Rep.* 2008; 57(27):737–41. PMID 18614991.
40. WHO Expert Committee on Oral Health Status and Fluoride Use. Fluorides and oral health [PDF]. 1994.
41. National Health and Medical Research Council (Australia). *A systematic review of the efficacy and safety of fluoridation* [PDF]. 2007 [Retrieved 2009-10-13]. ISBN 1-86496-415-4. Summary: Yeung CA. A systematic review of the efficacy and safety of fluoridation. *Evid Based Dent.* 2008; 9(2):39–43. doi:10.1038/sj.ebd.6400578. PMID 18584000. Lay summary: *NHMRC*, 2007.

42. U.S. Department of Health & Human Services. HHS and EPA announce new scientific assessments and actions on fluoride; 2011.
43. Fawell J, Bailey K, Chilton J, Dahi E, Fewtrell L, Magara Y. *Fluoride in Drinking-water* [PDF]. World Health Organization; 2006. ISBN 92-4-156319-2. Environmental occurrence, geochemistry and exposure. p. 5–27.
44. Ozsvath DL. Fluoride and environmental health: a review. *Rev Environ Sci Biotechnol.* 2009; 8(1):59–79. doi:10.1007/s11157-008-9136-9.
45. Taricska JR, Wang LK, Hung YT, Li KH. Fluoridation and defluoridation. In: Wang LK, Hung YT, Shammas NK, editors. *Advanced Physicochemical Treatment Processes*. Humana Press; 2006. (Handbook of Environmental Engineering 4). doi:10.1007/978-1-59745-029-4_9. ISBN 978-1-59745-029-4. p. 293–315.
46. Hobson WL, Knochel ML, Byington CL, Young PC, Hoff CJ, Buchi KF. Bottled, filtered, and tap water use in Latino and non-Latino children. *Arch Pediatr Adolesc Med.* 2007; 161(5):457–61. doi:10.1001/archpedi.161.5.457. PMID 17485621.
47. Lalumandier JA, Ayers LW. Fluoride and bacterial content of bottled water vs tap water. *Arch Fam Med.* 2000; 9(3):246–50. doi:10.1001/archfami.9.3.246. PMID 10728111.
48. Grec RHdC, de Moura PG, Pessan JP, Ramires I, Costa B, Buzalaf MAR. Fluoride concentration in bottled water on the market in the municipality of São Paulo. *Rev Saúde Pública.* 2008; 42(1):154–7. doi:10.1590/S0034-89102008000100022. PMID 18200355.
49. Featherstone JD. Dental caries: a dynamic disease process. *Aust Dent J.* 2008; 53(3):286–91. doi:10.1111/j.1834-7819.2008.00064.x. PMID 18782377.
50. Cury JA, Tenuta LM. How to maintain a cariostatic fluoride concentration in the oral environment. *Adv Dent Res.* 2008; 20(1):13–6. doi:10.1177/15440737080-2000104. PMID 18694871.
51. Pizzo G, Piscopo MR, Pizzo I, Giuliana G. Community water fluoridation and caries prevention: a critical review. *Clin Oral Investig.* 2007; 11(3):189–93. doi:10.1007/s00784-007-0111-6. PMID 17333303.
52. Aoba T, Fejerskov O. Dental fluorosis: chemistry and biology. *Crit Rev Oral Biol Med.* 2002; 13(2):155–70. doi:10.1177/154411130201300206. PMID 12097358.
53. Hellwig E, Lennon AM. Systemic versus topical fluoride [PDF]. *Caries Res.* 2004; 38(3):258–62. doi:10.1159/000077764. PMID 15153698.
54. Tinanoff N. Uses of fluoride. In: Berg JH, Slayton RL, editors. *Early Childhood Oral Health.* Wiley-Blackwell; 2009. ISBN 978-0-8138-2416-1. p. 92–109.

55. Koo H. Strategies to enhance the biological effects of fluoride on dental biofilms. *Adv Dent Res*. 2008; 20(1): 17–21. doi:10.1177/15440737080-2000105. PMID 18694872.
56. Marquis RE, Clock SA, Mota-Meira M. Fluoride and organic weak acids as modulators of microbial physiology. *FEMS Microbiol Rev*. 2003; 26(5):493–510. doi:10.1016/S0168-6445(02)00143-2. PMID 12586392.
57. Fawell J, Bailey K, Chilton J, Dahi E, Fewtrell L, Magara Y. *Fluoride in Drinking-water* [PDF]. World Health Organization; 2006. ISBN 92-4-156319-2. Human health effects. p. 29–36.
58. Sheiham A. Dietary effects on dental diseases [PDF]. *Public Health Nutr*. 2001; 4(2B):569–91. doi:10.1079/PHN2001142. PMID 11683551.
59. Institute of Medicine. *Dietary Reference Intakes for Calcium, Phosphorus, Magnesium, Vitamin D, and Fluoride*. National Academy Press; 1997. ISBN 0-309-06350-7. Fluoride. p. 288–313.
60. Parnell C, Whelton H, O'Mullane D. Water fluoridation. *Eur Arch Paediatr Dent*. 2009; 10(3):141–8. PMID 19772843.
61. McDonagh M, Whiting P, Bradley M *et al.* A systematic review of public water fluoridation [PDF]; 2000. Report website: NHS Centre for Reviews and Dissemination. Fluoridation of drinking water: a systematic review of its efficacy and safety; 2000 [Retrieved 2009-05-26]. Authors' summary: McDonagh MS, Whiting PF, Wilson PM *et al.*. Systematic review of water fluoridation [PDF]. *BMJ*. 2000; 321(7265):855–9. doi:10.1136/bmj.321.7265.855. PMID 11021861. PMC 27492. Authors' commentary: Treasure ET, Chestnutt IG, Whiting P, McDonagh M, Wilson P, Kleijnen J. The York review—a systematic review of public water fluoridation: a commentary. *Br Dent J*. 2002; 192(9):495–7. doi:10.1038/sj.bdj.4801410a. PMID 12047121.
62. Worthington H, Clarkson J. The evidence base for topical fluorides. *Community Dent Health*. 2003; 20(2):74–6. PMID 12914024.
63. Griffin SO, Regnier E, Griffin PM, Huntley V. Effectiveness of fluoride in preventing caries in adults. *J Dent Res*. 2007; 86(5):410–5. doi:10.1177/154405910708600504. PMID 17452559. Summary: Yeung CA. Fluoride prevents caries among adults of all ages. *Evid Based Dent*. 2007;8(3):72–3. doi:10.1038/sj.ebd.6400506. PMID 17891121.
64. Truman BI, Gooch BF, Sulemana I *et al.*. Reviews of evidence on interventions to prevent dental caries, oral and pharyngeal cancers, and sports-related craniofacial

injuries [PDF]. *Am J Prev Med.* 2002; 23(1 Suppl):21–54. doi:10.1016/S0749-3797(02)00449-X. PMID 12091093.

65. Burt BA, Tomar SL. Changing the face of America: water fluoridation and oral health. In: Ward JW, Warren C. *Silent Victories: The History and Practice of Public Health in Twentieth-century America.* Oxford University Press; 2007. ISBN 0-19-515069-4. p. 307–22.
66. Hausen HW. Fluoridation, fractures, and teeth. *BMJ.* 2000; 321(7265):844–5. doi:10.1136/bmj.321.7265.844. PMID 11021844.
67. Kumar JV. Is water fluoridation still necessary?. *Adv Dent Res.* 2008; 20(1):8–12. doi:10.1177/154407370802000103. PMID 18694870.
68. Alvarez JA, Rezende KMPC, Marocho SMS, Alves FBT, Celiberti P, Ciamponi AL. Dental fluorosis: exposure, prevention and management [PDF]. *Med Oral Patol Oral Cir Bucal.* 2009;14(2):E103–7. PMID 19179949.
69. Hujoel PP, Zina LG, Moimaz SAS, Cunha-Cruz J. Infant formula and enamel fluorosis: a systematic review. *J Am Dent Assoc.* 2009; 140(7):841–54. PMID 19571048.
70. Fawell J, Bailey K, Chilton J, Dahi E, Fewtrell L, Magara Y. *Fluoride in Drinking-water* [PDF]. World Health Organization; 2006. ISBN 92-4-156319-2. Guidelines and standards. p. 37–9.
71. Balbus JM, Lang ME. Is the water safe for my baby?. *Pediatr Clin North Am.* 2001;48(5):1129–52, viii. doi:10.1016/S0031-3955(05)70365-5. PMID 11579665.
72. Asheboro notifies residents of over-fluoridation of water. 2010-06-29. Fox 8.
73. Pollick HF. Water fluoridation and the environment: current perspective in the United States [PDF]. *Int J Occup Environ Health.* 2004; 10(3):343–50. PMID 15473093.
74. Macek MD, Matte TD, Sinks T, Malvitz DM. Blood lead concentrations in children and method of water fluoridation in the United States, 1988–1994. *Environ Health Perspect.* 2006; 114(1): 130–4. doi:10.1289/ ehp.8319. PMID 16393670. PMC 1332668.
75. Anusavice KJ. Present and future approaches for the control of caries. *J Dent Educ.* 2005; 69(5):538–54. PMID 15897335.
76. Jones S, Burt BA, Petersen PE, Lennon MA. The effective use of fluorides in public health. *Bull World Health Organ.* 2005; 83(9):670–6. PMID 16211158. PMC 2626340.

77. Milgrom P, Reisine S. Oral health in the United States: the post-fluoride generation. *Annu Rev Public Health.* 2000; 21: 403–36. doi:10.1146/annurev.publhealth.21.1.403. PMID 10884959.
78. Goldman AS, Yee R, Holmgren CJ, Benzian H. Global affordability of fluoride toothpaste. *Global Health.* 2008; 4: 7. doi:10.1186/1744-8603-4-7. PMID 18554382. PMC 2443131.
79. Bánóczy J, Rugg-Gunn AJ. Milk—a vehicle for fluorides: a review [PDF]. *Rev Clin Pesq Odontol.* 2006 [Retrieved 2009-01-03];2(5–6):415–26.
80. Yeung CA, Hitchings JL, Macfarlane TV, Threlfall AG, Tickle M, Glenny AM. Fluoridated milk for preventing dental caries. *Cochrane Database Syst Rev.* 2005; (3): CD003876. doi:10.1002/14651858. CD003876.pub2. PMID 16034911.
81. Bruvo M, Ekstrand K, Arvin E *et al.* Optimal drinking water composition for caries control in populations. *J Dent Res.* 2008; 87(4): 340–3. doi: 10.1177/154405910808700407. PMID 18362315.
82. Zero DT. Are sugar substitutes also anticariogenic?. *J Am Dent Assoc.* 2008;139(Suppl 2):9S–10S. PMID 18460675.
83. Whelton H. Beyond water fluoridation; the emergence of functional foods for oral health. *Community Dent Health.* 2009; 26(4): 194–5. doi: 10.1922/CDH_2611 Whelton02. PMID 20088215.
84. Consumer Price Index (estimate) 1800–2014. Federal Reserve Bank of Minneapolis. Retrieved February 27, 2014.
85. Marthaler TM, Petersen PE. Salt fluoridation—an alternative in automatic prevention of dental caries [PDF]. *Int Dent J.* 2005; 55(6):351–8. PMID 16379137.
86. Ho K, Neidell M. Equilibrium effects of public goods: the impact of community water fluoridation on dentists [PDF]. 2009 [Retrieved 2009-10-13].
87. The British Fluoridation Society; The UK Public Health Association; The British Dental Association; The Faculty of Public Health. *One in a Million: The facts about water fluoridation.* 2nd ed. Manchester: British Fluoridation Society; 2004. ISBN 0-9547684-0-X. The ethics of water fluoridation [PDF]. p. 88–92.
88. ADA Council on Access, Prevention and Interprofessional Relations. American Dental Association. National and international organizations that recognize the public health benefits of community water fluoridation for preventing dental decay; 2005 [archived 2008-06-07; Retrieved 2008-12-22].
89. Division of Oral Health, National Center for Chronic Disease Prevention and Health Promotion, CDC. Achievements in public health, 1900–1999: Fluoridation of drinking water to prevent dental caries. *MMWR Morb Mortal Wkly Rep.*

1999;48(41):933–40. Contains H. Trendley Dean, D.D.S. Reprinted in: *JAMA*. 2000; 283(10): 1283–6. doi:10.1001/jama. 283.10.1283. PMID 10714718.

90. CDC. Ten great public health achievements—United States, 1900–1999. *MMWR Morb Mortal Wkly Rep*. 1999;48(12):241–3. PMID 10220250. Reprinted in: *JAMA*. 1999; 281(16):1481. doi:10.1001/jama.281.16.1481. PMID 10227303.
91. Carmona RH. U.S. Public Health Service. Surgeon General's statement on community water fluoridation [PDF]; 2004-07-28 [Retrieved 2008-12-22].
92. American Public Health Association. Community water fluoridation in the United States; 2008 [Retrieved 2009-03-09].
93. European Academy Of Paediatric Dentistry. Guidelines on the use of fluoride in children: an EAPD policy document [PDF]. *Eur Arch Paediatr Dent*. 2009; 10(3):129–35. PMID 19772841.
94. Australian Dental Association. Community oral health promotion: fluoride use [PDF]; 2005 [Retrieved 2009-10-13].
95. Canadian Dental Association. CDA position on use of fluorides in caries prevention [PDF]; 2008 [Retrieved 2009-01-15].
96. ADA Council on Access, Prevention and Interprofessional Relations. American Dental Association. Fluoridation facts [PDF]; 2005 [archived 2008-07-23; Retrieved 2008-12-22].
97. Armfield JM. When public action undermines public health: a critical examination of antifluoridationist literature. *Aust New Zealand Health Policy*. 2007; 4:25. doi:10.1186/1743-8462-4-25. PMID 18067684. PMC 2222595.
98. Carstairs C, Elder R. Expertise, health, and popular opinion: debating water fluoridation, 1945–80. *Can Hist Rev*. 2008; 89(3):345–71. doi: 10.3138/chr. 89.3.345.
99. National Institute of Dental and Craniofacial Research. The story of fluoridation; 2008-12-20 [Retrieved 2010-02-06].
100. Fagin D. Second thoughts about fluoride. *Sci Am*. 2008; 298(1):74–81. doi:10.1038/scientificamerican0108-74. PMID 18225698.
101. National Research Council. *Fluoride in Drinking Water: A Scientific Review of EPA's Standards*. Washington, DC: National Academies Press; 2006. ISBN 0-309-10128-X. Lay summary: *NRC*, 2006.
102. Reilly GA. The task is a political one: the promotion of fluoridation. In: Ward JW, Warren C. *Silent Victories: The History and Practice of Public Health in Twentieth-century America*. Oxford University Press; 2007. ISBN 0-19-515069-4. p. 323–42.

103. Libertarian Party. Consumer protection [Retrieved June 28, 2010].
104. Freeze RA, Lehr JH. *The Fluoride Wars: How a Modest Public Health Measure Became America's Longest-Running Political Melodrama.* Wiley; 2009. ISBN 978-0-470-44833-5. p. 62.
105. Nordlinger J. Water fights: believe it or not, the fluoridation war still rages—with a twist you may like. *Natl Rev.* 2003-06-30.
106. *The Fluoride Wars*. ISBN 0-470-44833-4. Fluoride and health. p. 219–54.
107. Griffin M, Shickle D, Moran N. European citizens' opinions on water fluoridation. *Community Dent Oral Epidemiol.* 2008; 36(2):95–102. doi:10.1111/j.16000528.2007.00373.x. PMID 18333872.
108. Armfield JM, Akers HF. Risk perception and water fluoridation support and opposition in Australia. *J Public Health Dent.* 2009; 70(1):58–66. doi:10.1111/j.1752-7325.2009.00144.x. PMID 19694932.
109. Dixon S, Shackley P. Estimating the benefits of community water fluoridation using the willingness-to-pay technique: results of a pilot study. *Community Dent Oral Epidemiol.* 1999; 27(2):124–9. doi:10.1111/j.1600-0528.1999.tb02001.x. PMID 10226722.
110. Calman K. Beyond the 'nanny state': stewardship and public health. *Public Health.* 2009; 123(1):e6–e10. doi:10.1016/j.puhe.2008.10.025. PMID 19135693. Lay summary: *Nuffield Council on Bioethics*, 2007-11-13.
111. Martin B. The sociology of the fluoridation controversy: a reexamination. *Sociol Q.* 1989; 30(1):59–76. doi:10.1111/j.1533-8525.1989.tb01511.x.
112. Freeze RA, Lehr JH. *The Fluoride Wars: How a Modest Public Health Measure Became America's Longest-Running Political Melodrama.* Wiley; 2009. ISBN 978-0-470-44833-5. Fluorophobia. p. 127–69.
113. Newbrun E. The fluoridation war: a scientific dispute or a religious argument?. *J Public Health Dent.* 1996; 56(5 Spec No):246–52. doi:10.1111/j.1752-7325.1996.tb02447.x. PMID 9034969.

# Chapter-5 Fluoride toxicity

## 5. Fluoride toxicity

### 5.1 History of Fluoride toxicity:

Danish researcher Kaj Roholm published *Fluorine Intoxication* in 1937, which was praised in a 1938 review by dental researcher H. Trendley Dean as "probably the outstanding contribution to the literature of fluorine".[1] Since that time, the fluoridation of public water has been widely implemented and has been hailed as one of the top medical achievements of the 20th Century.[2] The effects of fluoride-rich ground water became recognized in the 1990s.[3]

In high concentrations, soluble fluoride salts are toxic and skin or eye contact with high concentrations of many fluoride salts is dangerous. Referring to a common salt of fluoride, sodium fluoride (NaF), the lethal dose for most adult humans is estimated at 5 to 10 g (which is equivalent to 32 to 64 mg/kg elemental fluoride/kg body weight).[4][5][6] Ingestion of fluoride can produce gastrointestinal discomfort at doses at least 15 to 20 times lower (0.2–0.3 mg/kg) than lethal doses.[7] Although helpful for dental health in low dosage, chronic exposure to fluoride in large amounts interferes with bone formation. In this way, the greatest examples of fluoride poisoning arise from fluoride-rich ground water.[3]

### 5.2 Regulatory background

For optimal dental health, the World Health Organization recommends a level of fluoride from 0.5 to 1.0 mg/L (milligrams per litre), depending on climate.[8] Adverse effects are possible at fluoride levels far above this recommended dosage. The United States Health and Human Services Department recommends 0.7 milligrams of fluoride per liter of water – the lower limit of the current recommended range of 0.7 to 1.2 milligrams.[9]

### 5.2.1 Acute toxicity

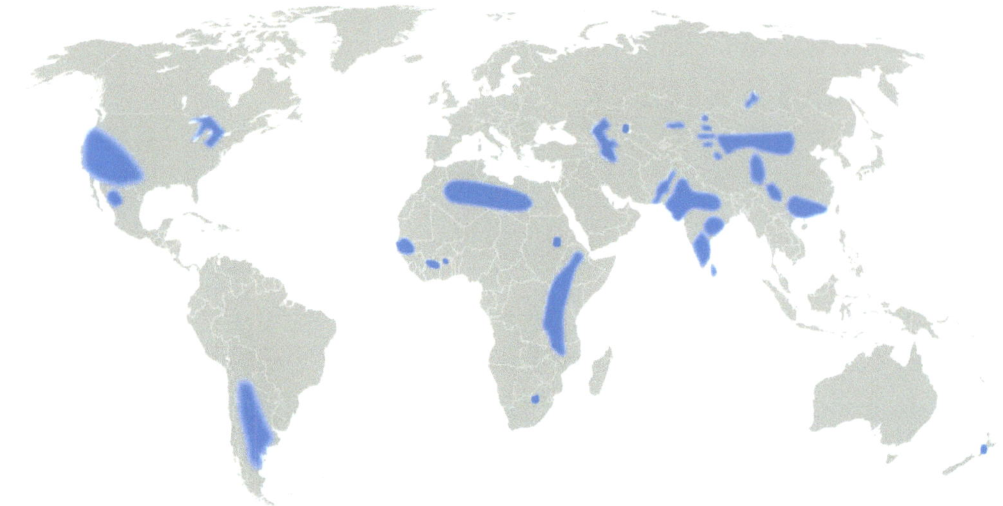

*Geographical areas associated with groundwater having over 1.5 mg/L of naturally occurring fluoride, which is above recommended levels.*[10]

In India an estimated 60 million people have been poisoned by well water contaminated by excessive fluoride, which is dissolved from the granite rocks. The effects are particularly evident in the bone deformations of children. Similar or larger problems are anticipated in other countries including China, Uzbekistan, and Ethiopia.[3]

Historically, most cases of acute fluoride toxicity have followed accidental ingestion of sodium fluoride based insecticides or rodenticides.[11] Currently, in advanced countries, most cases of fluoride exposure are due to the ingestion of dental fluoride products.[12] Although exposure to these products does not often cause toxicity, in one study 30% of children exposed to fluoride dental products developed mild symptoms.[12] Other sources include glass-etching or chrome-cleaning agents like ammonium bifluoride or hydrofluoric acid,[13][14] industrial exposure to fluxes used to promote the flow of a molten metal on a solid surface, volcanic ejecta (for example, in cattle grazing after an 1845–1846 eruption of Hekla and the 1783–1784 flood basalt eruption of Laki), and metal cleaners. Malfunction of water fluoridation equipment has happened several times, including a notable incident in Alaska.[7]

#### *(a) Organofluorine compounds*

Organofluorine compounds only rarely release $F^-$ under biological conditions and thus are rarely sources of fluoride poisoning. In order for fluoride poisoning to occur, a compound must release fluoride ($F^-$) ions. Whereas most organofluorine compounds may not release $F^-$ because of the strength of the carbon–fluorine bond and its tendency to strengthen as more fluorine atoms are added to a carbon, some compounds do, such as methoxyflurane.

The fluorine atom is pervasive in drugs, e.g. Prozac and many other substances such as freon, Teflon, and blood serum (PFOS, PFOA, and PFNA).

Children may experience gastrointestinal distress upon ingesting sufficient amounts of flavored toothpaste. Between 1990 and 1994, over 628 people, mostly children, were treated after ingesting too much fluoride-containing toothpaste. "While the outcomes were generally not serious," gastrointestinal symptoms appear to be the most common problem reported.[15]

### 5.3 Effect of fluoride on different organ:

Here we discuss about the effect of excess fluoride on the different organ of human in detail as below.

#### 5.3.1 Brain

A meta analysis conducted on 27 epidemiological studies (most from China), concluded that exposure to "high levels" of fluoride in childhood was associated with a reduction in IQ. The authors note that this research is not applicable to the safety of artificial water fluoridation because the adverse effects on IQ was found with fluoride levels that were much higher than typically found in artificially fluoridated water.[16] However, they conclude that more research is required to assess the adverse effects on children's neurodevelopment.[17][18] The meta analysis has been criticized for failing to account for confounding factors. For example, in some of the studies fluoride exposure came from the burning of high fluoride content coal, and used a control group from an area in which wood was used as fuel.[19] A more recent study followed individuals over 38 years to see if fluoride exposure affected IQ and they concluded there was no link between fluoride exposure and IQ, or fluoride as a neurotoxin.[20]

#### 5.3.2 Bones

Whilst fluoridated water is associated with decreased levels of fractures in a population,[21] toxic levels of fluoride have been associated with a weakening of bones and an increase in hip and wrist fractures. The U.S. National Research Council concludes that fractures with fluoride levels 1–4 mg/L, suggesting a dose-response relationship, but states that there is "suggestive but inadequate for drawing firm conclusions about the risk or safety of exposures at [2 mg/L]".[22]:170

Consumption of fluoride at levels beyond those used in fluoridated water for a long period of time causes skeletal fluorosis. In some areas, particularly the Asian subcontinent, skeletal fluorosis is endemic. It is known to cause irritable-bowel symptoms and joint pain.

Early stages are not clinically obvious, and may be misdiagnosed as (seronegative) rheumatoid arthritis or ankylosing spondylitis.[23]

### 5.3.3 Kidney

Fluoride induced nephrotoxicity is kidney injury due to toxic levels of serum fluoride, commonly due to release of fluoride from fluorine-containing drugs, such as methoxyflurane.[24][25][26]

Within the recommended dose, no effects are expected, but chronic ingestion in excess of 12 mg/day are expected to cause adverse effects, and an intake that high is possible when fluoride levels are around 4 mg/L.[22]:281 Those with impaired kidney function are more susceptible to adverse effects.[22]:292

The kidney injury is characterised by failure to concentrate urine, leading to polyuria, and subsequent dehydration with hypernatremia and hyperosmolarity. Inorganic fluoride inhibits adenylate cyclase activity required for antidiuretic hormone effect on the distal convoluted tubule of the kidney. Fluoride also stimulates intrarenal vasodilation, leading to increased medullary blood flow, which interferes with the counter current mechanism in the kidney required for concentration of urine.

Fluoride induced nephrotoxicity is dose dependent, typically requiring serum fluoride levels exceeding 50 micromoles per liter (about 1 ppm) to cause clinically significant renal dysfunction,[27] which is likely when the dose of methoxyflurane exceeds 2.5 MAC hours.[28][29] (Note: "MAC hour" is the multiple of the minimum alveolar concentration (MAC) of the anesthetic used times the number of hours the drug is administered, a measure of the dosage of inhaled anesthetics.) Elimination of fluoride depends on glomerular filtration rate. Thus, patients with renal insufficiency will maintain serum fluoride for longer period of time, leading to increased risk of fluoride induced nephrotoxicity.

### 5.3.4 Teeth

The only generally accepted adverse effect of fluoride at levels used for water fluoridation is dental fluorosis, which can alter the appearance of children's teeth during tooth development; this is mostly mild and usually only an aesthetic concern. Compared to unfluoridated water, fluoridation to 1 mg/L is estimated to cause fluorosis in one of every 6 people (range 4–21), and to cause fluorosis of aesthetic concern in one of every 22 people (range 13.6–∞).[21]

### 5.3.5 Thyroid

Fluoride's suppressive effect on the thyroid is more severe when iodine is deficient, and fluoride is associated with lower levels of iodine.[clarification needed][30] Thyroid effects in humans were associated with fluoride levels 0.05–0.13 mg/kg/day when iodine intake was adequate and 0.01–0.03 mg/kg/day when iodine intake was inadequate.[22]:263 Its mechanisms and effects on the endocrine system remain unclear.[22]:266

## 5.4 Mechanism of fluoride

Like most soluble materials, fluoride compounds are readily absorbed by the stomach and intestines, and excreted through the urine. Urine tests have been used to ascertain rates of excretion in order to set upper limits in exposure to fluoride compounds and associated detrimental health effects.[31] Ingested fluoride initially acts locally on the intestinal mucosa, where it forms hydrofluoric acid in the stomach.

The NRC report stated that "many of the untoward effects of fluoride are due to the formation of AlFx [aluminum fluoride] complexes".[22]:219 This topic has been identified previously as cause for concern.[30] The NRC noted that rats administered fluoride had twice as much aluminum in their brains.[22]:212 When water (1 ppm fluoride) is boiled in aluminum cookware more aluminum is leached and more aluminum fluoride complexes are formed. However, an epidemiological study found that a high-fluoride area had one-fifth the Alzheimer's that a low-fluoride area had,[32] and a 2002 study found that fluoride increased the urinary excretion of aluminum.[33]

## 5.5 References


1. Dean T.H. (1938). **Fluorine Intoxication**. *Am J Public Health Nations Health* 28: 1008–1009. Free full text.
2. Division of Oral Health, National Center for Chronic Disease Prevention and Health Promotion, CDC. Achievements in public health, 1900–1999: Fluoridation of drinking water to prevent dental caries. *MMWR Morb Mortal Wkly Rep*. 1999; 48(41):933–40. Contains H. Trendley Dean, D.D.S. Reprinted in: *JAMA*. 2000; 283(10):1283–6. doi:10.1001/ jama. 283.10.1283. PMID 10714718.
3. Pearce, Fred (2006). *When the Rivers Run Dry: Journeys Into the Heart of the World's Water Crisis*. Toronto: Key Porter. ISBN 978-1-55263-741-8.
4. Gosselin, RE; Smith RP; Hodge HC (1984). *Clinical toxicology of commercial products*. Baltimore (MD): Williams & Wilkins. pp. III–185–93. ISBN 0-683-03632-7.

5. Baselt, RC (2008). *Disposition of toxic drugs and chemicals in man*. Foster City (CA): Biomedical Publications. pp. 636–40. ISBN 978-0-9626523-7-0.
6. IPCS (2002). *Environmental health criteria 227 (Fluoride)*. Geneva: International Programme on Chemical Safety, World Health Organization. p. 100. ISBN 92-4-157227-2.
7. Bradford D. Gessner; Michael Beller; John P. Middaugh; Gary M. Whitford (13 January 1994). "Acute fluoride poisoning from a public water system". *New England Journal of Medicine* **330** (2): 95–99. doi:10.1056/NEJM199401-133300203. PMID 8259189.
8. WHO Expert Committee on Oral Health Status and Fluoride Use. Fluorides and oral health [PDF]. 1994.
9. http://www.reuters.com/article/2011/01/08/us-usa-fluoride-idUSTRE7064CM20110108
10. National Health and Medical Research Council (Australia). *A systematic review of the efficacy and safety of fluoridation* [PDF]. 2007 [Retrieved 2009-10-13]. ISBN 1-86496-415-4. Summary: Yeung CA. A systematic review of the efficacy and safety of fluoridation. *Evid Based Dent*. 2008;9(2):39–43. doi:10.1038/sj.ebd.6400578. PMID 18584000. Lay summary: *NHMRC*, 2007.
11. Nochimson G. (2008). Toxicity, Fluoride. eMedicine. Retrieved 2008-12-28.
12. Augenstein WL, Spoerke DG, Kulig KW, et al. (November 1991). "Fluoride ingestion in children: a review of 87 cases". *Pediatrics* **88** (5): 907–12. PMID 1945630.
13. Wu ML, Deng JF, Fan JS (November 2010). "Survival after hypocalcemia, hypomagnesemia, hypokalemia and cardiac arrest following mild hydrofluoric acid burn". *Clinical Toxicology (Philadelphia, Pa.)* **48** (9): 953–5. doi:10.3109/15563650.2010.533676. PMID 21171855.
14. Klasaer AE, Scalzo AJ, Blume C, Johnson P, Thompson MW (December 1996). "Marked hypocalcemia and ventricular fibrillation in two pediatric patients exposed to a fluoride-containing wheel cleaner". *Annals of Emergency Medicine* **28** (6): 713–8. doi:10.1016/S0196-0644(96)70097-5. PMID 8953969.
15. Jay D. Shulman; Linda M. Wells (1997). "Acute Fluoride Toxicity from Ingesting Home-use Dental Products in Children, Birth to 6 Years of Age". *Journal of Public Health Dentistry* **57** (3): 150–158. doi:10.1111/j.1752-7325.1997.tb02966.x. PMID 9383753.
16. Lefler, Dion (11 September 2012). "Harvard scientists: Data on fluoride, IQ not applicable in U.S.". *The Wichita Eagle*. Retrieved 5 March 2014.

17. Anna L. Choi, Guifan Sun, Ying Zhang, and Philippe Grandjean (2012). "Developmental fluoride neurotoxicity: a systematic review and meta-analysis.". *Environ. Health Perspect.* (USA: National Center for Biotechnology Information) **120** (10): 1362–8. doi:10.1289/ehp.1104912. PMC 3491930. PMID 22820538.
18. "Neurobehavioural effects of developmental toxicity". *The Lancet Neurology, Volume 13, Issue 3, Pages 330 - 338, March 2014*. USA: Elsevier Ltd.
19. Sabour S, Ghorbani Z (March 2013). "Developmental fluoride neurotoxicity: clinical importance versus statistical significance". *Environ. Health Perspect.* **121** (3): A70. doi:10.1289/ehp.1206192. PMC 3621182. PMID 23455234.
20. Broadbent, Jonathan M.; Thomson, W. Murray; Ramrakha, Sandhya; Moffitt, Terrie E.; Zeng, Jiaxu; Foster Page, Lyndie A.; Poulton, Richie (15 May 2014). "Community Water Fluoridation and Intelligence: Prospective Study in New Zealand". *American Journal of Public Health*: e1–e5. doi:10.2105/AJPH. 2013.301857. Retrieved 4 July 2014.
21. McDonagh MS et al. Systematic review of water fluoridation BMJ. 2000 Oct 7;321(7265):855-9. doi:10.1136/bmj.321.7265.855 PMID 11021861
22. National Research Council (2006). *Fluoride in Drinking Water: A Scientific Review of EPA's Standards*. Washington, DC: National Academies Press. ISBN 0-309-10128-X. Lay summary – *NRC* (September 24, 2008).. See also CDC's statement on this report.
23. Gupta R, Kumar AN, Bandhu S, Gupta S (2007). "Skeletal fluorosis mimicking seronegative arthritis". *Scand. J. Rheumatol.* **36** (2): 154–5. doi:10.1080/03009740600759845. PMID 17476625.
24. Cousins MJ, Skowronski G, Plummer JL. Anaesthesia and the kidney. Anaesth Intensive Care. 1983 Nov; 11(4):292-320.
25. Baden JM, Rice SA, Mazze RI. Deuterated methoxyflurane anesthesia and renal function in Fischer 344 rats. *Anesthesiology*. 1982 Mar; 56(3): 203-6.
26. Mazze RI. Methoxyflurane nephropathy. *Environ Health Perspect*. 1976 Jun; 15: 111-9.
27. Cousins MJ, Greenstein LR, Hitt BA, Mazze RI. Metabolism and renal effect of enflurane in men. *Anesthesiology* 1976; 44: 44-53.
28. VanDyke R. Biotransformation of volatile anesthetics with special emphasis on the role of metabolism in the toxicity of anesthetics. Can Anaesth Soc J 1973;20:21-33.
29. White AE, Stevens WC, Eger EI II, Mazze RI, Hitt BA. Enflurane and methoxyflurane metabolism at anesthetic and subanesthetic concentrations. Anesth Analg 1979; 58: 221-4/

30. Strunecká A, Strunecký O, Patocka J (2002). "Fluoride plus aluminum: useful tools in laboratory investigations, but messengers of false information". *Physiol Res* **51** (6): 557–64. PMID 12511178.
31. Baez, Ramon J.; Baez, Martha X.; Marthaler, Thomas M. (2000). "Urinary fluoride excretion by children 4–6 years old in a south Texas community". *Revista Panamericana de Salud Pública/Pan American Journal of Public Health* **7** (4): 242–248. doi:10.1590/s1020498920000-00400005.
32. Li L (2003). "The biochemistry and physiology of metallic fluoride: action, mechanism, and implications". *Crit. Rev. Oral Biol. Med.* **14** (2): 100–14. doi:10.1177/154411130301400204. PMID 12764073.
33. Chiba J, Kusumoto M, Shirai S, Ikawa K, Sakamoto S (March 2002). "The influence of fluoride ingestion on urinary aluminum excretion in humans". *Tohoku J. Exp. Med.* **196** (3): 139–49. doi:10.1620/tjem. 196.139. PMID 12002270. Free full-text.

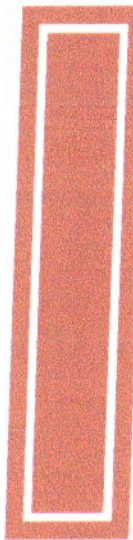

# Chapter-6

# Fluoride Test and Fluoride Leveis in Human Plasma and Brest Milk

## 6. Fluoride Test and Fluoride Levels in Human Plasma and Breast Milk

### 6.1 Introduction:

Since various forms of fluorides have met wide acceptance for use in the prevention of dental caries, the metabolism of fluoride is of considerable interest.[1–3] The human organism is exposed to fluoride in a number of ways. Ingestion of fluoride is accomplished through various foods; drinking water and fluoride containing products comprising dentifrices, mouth rinses, tablets, drops, etc.[4–6] Hard tissues are known to be the major sites of fluoride accumulation in the human body. Approximately 99% of the total body burden of fluoride is retained in bones and teeth, with the remainder distributed in highly-vascularized soft tissues.[1,5] Kidneys are the primary route for the removal of fluoride from the body. Other routes of fluoride excretion are sweat, feces, saliva and breast milk of lactating mothers.[4,5,7]

Breast milk is the major dietary intake of infants in the early stage of life. The level of fluoride in breast milk plays an important role as a fluoride supplement to the infant.[8] Conversely, the concentration of fluoride could be deemed critical regarding the potential dental fluorosis that may result from high concentrations of dietary fluoride.[9,10]

Plasma is the biological fluid into which fluoride must pass for its distribution elsewhere in the body as well as its elimination from the body. For these reasons, plasma is often referred to as the central compartment of the body.[6] Factors that include fluoride intake from various sources may affect plasma fluoride levels, and thus fluoride content of breast milk.

The aim of this pilot study was to determine the fluoride levels of breast milk and plasma of lactating mothers and the correlation between breast milk and plasma fluoride levels in mothers who regularly consume drinking water with low levels of fluoride.

### 6.2 Fluoride test:

#### 6.2.1 Urine Test For Fluoride

The least expensive lab test is a 24 hour urine test in which a person collects their urine for a 24 hour period. The fluoride concentration in the sample is measured using an ion specific electrode meter. A result of more than 1 ppm (1 mg/L) means there is fluoride overdose.

If a person's daily fluoride intake is low but their urine fluoride is high, this indicates chronic fluoride poisoning. The high urine fluoride level is caused by fluoride being released from bone into the blood.

Doctor's Data Lab in Chicago can run this test (www.doctorsdata.com) at a reasonable cost. The test results may not be definitive because urine can concentrate fluoride up to 50 times more than what is in the blood, plus part of what is in your blood is from daily fluoride intake, which varies.

If desired, the test can be repeated to show whether there is consistently elevated blood fluoride over time. This would also be able to show that avoiding fluoride results in reduced urine fluoride and by extension, reduced blood fluoride.

### 6.2.2 Blood Test for Fluoride

The most expensive yet accurate test is a fasting serum (blood) test, requiring a licensed medical professional to draw blood, and send it to a medical lab. It is almost impossible to request this test through normal channels such as a family doctor in any country where artificial fluoridation is done.

Doctors are taught that increased fluoride intake is beneficial, and elevated serum fluoride is not harmful until it is grossly high. They believe there is no point in testing for serum fluoride unless the person has signs of acute fluoride poisoning with evidence of high exposure such as a workplace chemical spill or overdose from fluorinated drug or dental product. But even then, the toxicity will be underestimated and even scoffed at.

In theory, a medical doctor should be able to order several related medical tests that strongly indicate chronically elevated fluoride in the blood.

1. Blood calcium level: Any emergency room doctor would know to check this when someone arrives with symptoms of low blood calcium that may accompany heart attack or stroke – or fluoride poisoning.
2. Urine test for increased bone collagen protein breakdown: The results of this test can find evidence of bone loss – or chronic fluoride poisoning.
3. Parathyroid Hormone (PTH): This blood test measures abnormal calcium metabolism – or chronic fluoride poisoning.
4. Thyroid function blood tests: These blood tests measure evidence of hypothyroid function (such as increased TSH, change in ratio between Free T4 and Free T3) – or chronic fluoride poisoning.

## 6.3 Material and methods:

One hundred twenty five mothers aged between 20–30 years old with hospitalized newborns due to icterus neonatorum were included in the study. Signed consent was obtained from the participants after explanations regarding the study protocol. The human ethic committee of Selcuk University Experimental Research Center (SUDAM) approved this study (Approval No:2004–034).

Besides being otherwise healthy, the primary selection criteria stipulated the absence of fluoride supplement consumption one month before delivery. The participants regularly consumed drinking water from the same city supply which has been previously shown to contain low levels of fluoride (approx. 0.3 ppm).[11] The mothers consumed a regular hospital diet.

Milk and plasma samples were collected from lactating mothers within 5 to 7 days after delivery. For milk samples, the breast was swabbed with cotton wool and distilled water before milk collection. The mother was instructed to press the breast gently to facilitate collection of 5 ml of milk into a polyethylene tube. At the same appointment, 5 ml of blood was obtained and transferred into a fluoride-free heparinized polyethylene tube. Thereafter, the plasma was separated from the blood by centrifugation for 3 min at 3500 g. Milk and plasma samples were further stored at −18°C until analyses. Before fluoride measurements, the samples were thawed at room temperature.

To determine fluoride concentrations, equal volumes of TISAB II buffer (Orion Research, U.S.A.) was added into the samples. All samples were homogenized using magnetic stirrers throughout the measurements. An ion-selective electrode (Model 96–09, Orion Research, USA) was used in conjunction with a Model EA 910 ion analyzer (Orion Research, USA) to measure the fluoride concentrations of the breast milk and plasma samples.

Paired t test was used to determine the differences between fluoride concentration of breast milk and plasma. Pearson correlation analysis was used to assess any possible relationship between plasma and breast milk fluoride levels.[12]

## 6.4 Results:

The concentrations of fluoride in breast milk and plasma are presented in Table 1. The mean fluoride concentration of the plasma samples was 0.017±0.011 ppm (range 0.006–0.054 ppm). The mean fluoride concentration of the breast milk samples was 0.006±0.02 ppm (range 0.003–0.011 ppm).

| Table-1 Fluoride concentrations of breast milk and plasma. | | | |
|---|---|---|---|
| | **Min** | **Max** | **Mean±SD** |
| Plasma (n=125) | 0.006 | 0.054 | 0.017±0.011 |
| Breast milk (n=125) | 0.003 | 0.011 | 0.006±0.002 |

*Table 1 Fluoride concentrations of breast milk and plasma.*

Paired t test showed that the fluoride concentrations of plasma were significantly higher than those of the breast milk (P=.000). Pearson analysis revealed a significant correlation between the fluoride concentrations of breast milk and of plasma (P=.000). When a mother's plasma fluoride concentration was above (or below) the mean plasma fluoride level of the entire study group, the breast-milk fluoride levels were affected accordingly.

### 6.5 Discussion:

Several methods are used to determine fluoride levels in biologic tissues that include spectrophotometry,[13] gas chromatography,[14] capillary electrophoresis,[15] micro diffusion,[16] and ion analysis in conjunction with ion-selective electrodes.[17] As utilized in the present study, the most common procedure used to quantify free fluoride anion is the ion-selective electrode.[18]

The plasma fluoride concentration displays an increase along with fluoride intake. This increase is, however, attenuated due to distribution to the interstitial and intracellular fluid uptake by calcified tissues and renal excretion.[5] The literature contains a wide range (0.008–0.045 ppm) of reported normal plasma fluoride concentrations.[6,18] The diversity of values may have been due to the inclusion of fasting individuals as subjects in contrast to other studies employing non-fasting participants.[18] Certainly, other factors that include methodological variations as well as the fluoride levels of drinking-water consumed by subjects should have a strong impact on the reported values.[18] Li et al[19] reported a mean plasma fluoride concentration of 0.106±0.076 ppm in 127 subjects. In their study, the subjects were selected from a region with the drinking water fluoride concentra tions of 5.03 ppm. In the present study the mean plasma fluoride concentration was 0.017±0.011 ppm. Our finding corroborates those of Fejerskow et al[6] and World Health Organization (WHO).[18]

Breast milk possesses unique nutritional, biochemical, anti-infective and anti-allergic properties. As breast-fed infants obtain fluids almost exclusively from their mothers, breast milk represents an important way for delivering fluoride with certain levels to infants.[20] The level of fluoride in human milk has been a topic of investigation for many years. Medical literature contains a wide range for fluoride levels in breast milk. It is probable that problems with the analysis of fluoride have been contributory. According to the WHO,[18] the breast milk fluoride levels range from <0.002 to about 0.1 ppm, with most values being between 0.005–0.010 ppm. The mean breast milk fluoride concentrations obtained here in (0.006±0.002 ppm) are in line with the WHO.[18] Dabeka et al[8] showed that the concentration of fluoride in breast milk was related to the content of the drinking-water consumed by the mothers. The mean concentration of fluoride in breast milk obtained from 32 women consuming drinking water that contained < 0.16 ppm was 0.004 ppm, whereas breast milk obtained from 112 women consuming drinking water containing 1 ppm reportedly was 0.009 ppm.[8] Similar levels of fluoride concentrations of breast milk and colostrum (0.008 ppm) have been reported by Spak et al.[1] However, Spak et al[1]found no significant difference in breast milk fluoride concentrations of mothers living in areas with low and high drinking-water fluoride concentrations.

In the present study, the strict selection criteria which stipulated absence of recent use of fluoride supplements was a preventive measure to control variables that could interfere with the results. Additional limitations included selection of patients from a region with low drinking water fluoride levels (<0.3 ppm). In light of previous work,[8] however, it is apparent that the fluoride concentration of breast milk in mothers regularly consuming higher concentrations of fluoridated water is still within normal limits.

Theoretically, a limited transfer of fluoride from plasma to breast milk should occur.[21] The mechanism(s) responsible for the selective transfer of fluoride into breast tissue is obscure.[21] It is thought that a physiological plasma-milk barrier functions against to fluoride.[1,2,6] Despite high doses of supplementary fluoride administered to the mother, the child receives a maximum dose of only 0.2% of the mother's fluoride intake.[1,2] The results obtained in our study confirmed these conclusions. It should, however, be noted that the fluoride content transferred through breast milk is less than those present in cow's milk and in infant formulas, utilized as routine substitutes for breast milk. Rahul et al[3] found that fluoride concentrations of various commercially available infant milk formulations ranged from 1.95 ppm to 7.45 ppm and fluoride content of cow's milk samples was 0.12 ppm; values exceeding those of breast milk.

### 6.6 Effect of Fluoride level:

When the body gets a large single dose of fluoride or is frequently exposed to a steadily increased dose, it causes a drop in blood calcium and threatens the normal blood pH. This can happen from high intake of fluoridated water, a fluorinated drug, tea, industrial pollution or if kidneys are impaired and don't excrete fluoride well. This provokes an immediate release of PTH which rapidly causes bone-demolition cells called osteoclasts to break down some bone to release calcium. In this way calcium is used to replenish the mineral buffers that keep the blood at slightly alkaline pH.

Bone minerals are in a collagen matrix. There is a difference between these two issues:

1. Breakdown of bone collagen protein ratio in normal bone turnover;
2. Breakdown of bone collagen protein ratio in response to elevated blood fluoride/lowered calcium/increased PTH.

That is how the urine collagen protein test can be used to diagnose chronic fluoride poisoning, early stage skeletal fluorosis, and also the bone condition caused by fluoride poisoning that goes with kidney disease.

Ideally, this surge of PTH with resultant bone loss crisis would be directly followed by an increased activity of bone-rebuilding cells called osteoblasts putting back the minerals into bone. But with chronic fluoride poisoning that can't happen because the blood fluoride levels are chronically higher. So the result is progressively more bone loss with more release of stored fluoride, and less ability of kidneys to excrete it. A vicious cycle indeed.

**6.7 References:**

1. Spak CJ, Hardell LI, De Chateau P. Fluoride in human milk. Acta Paediatr Scand. 1983; 72:699–701.
2. Ekstrand J, Spak CJ, Falch J, Afseth J, Ulvestad H. Distribution of fluoride to human breast milk. Caries Res. 1984; 18:93–95.
3. Rahul P, Hedge AM, Munshi AK. Estimation of the fluoride in human breast milk, cow's milk and infant formulate. J Clin Pediatr Dent. 2003; 27:257–260.
4. Mellberg JR, Ripa LW, Leske GS. Fluoride in Preventive Dentistry. Chicago: Quintessence Publishing; 1983.
5. Whitford G. The Metabolism and Toxicity of Fluoride. 2. Switzerland: Karger; 1996.
6. Fejerskov O, Ekstrand J, Burt BA. Fluoride in Dentistry. 2. Copenhagen: Munsksgaard; 1996.
7. World Health Organization. Fluorides and oral health: Technical Report Series. Switzerland: 1994. p. 846.
8. Dabeka RW, Karpinski K, McKenzie A, Bajdik C. Survey of lead, cadmium and fluoride in human milk and correlation of levels with environmental and food factors. Food Chem Toxicol. 1986; 24:913–921.
9. Fejerskow O, Manji F, Baelum V. The nature and mechanism of dental fluorosis in man. J Dent Res. 1990; 69:699–700.
10. Ekstrand J, Fomon SJ, Ziegler EZ, Nelson SE. Fluoride pharmacokinetics in infancy. Pediatr Res. 1994; 35:157–163.
11. Sener Y, Koyuturk AE, Gokalp A. The fluoride levels of community water supplies in Konya city. The Journal of Hacettepe Faculty of Dentistry (in Turkish) 2003; 27:2–6.
12. Brunette DM. Critical Thinking: Understanding and evaluating dental research. 1. Chicago: Quintessence Publishing Co; 1996.

13. Greenhalgh R, Riley JP. The determination of fluoride in natural waters, with particular reference to sea water. Anal Chim Acta. 1961; 25:179–188.
14. Fresen JA, Cox FH, Witter MJ. The determination of fluoride in biological materials by means of gas chromatography. Pharm Weekbl. 1968; 103:909–914.
15. Yap AU, Tham SY, Zhu LY, Lee HK. Short-term fluoride release from various aesthetic restorative materials. Oper Dent. 2002; 27:259–265.
16. Kahama RW, Damen JJ, ten Cate JM. Enzymatic release of sequestered cows' milk fluoride for analysis by the hexamethyldisiloxane microdiffusion method. Analyst. 1997; 122:855–858.
17. Koparal E, Ertu grul F, Öztekin K. Fluoride levels in breast milk and infant foods. J Clin Pediatr Dent. 2000; 24:299–302.
18. World Health Organization. Environmental Health Criteria. Geneva: 2002. p. 227.
19. Li Y, Liang CK, Katz BP, Brizendine EJ, Stookey GK. Long-term exposure to fluoride in drinking water and sister-chromatid exchange frequency in human blood lymphocytes. J Dent Res. 1995; 74:1468–1474.
20. Latifah R, Razak IA. Fluoride levels in mother's milk. J Pedod. 1989;13:149–154.
21. Ekstrand J, Boreus LO, de Chateau P. No evidence of transfer of fluoride from plasma to breast milk. Br Med J. 1981; 283:761–762.

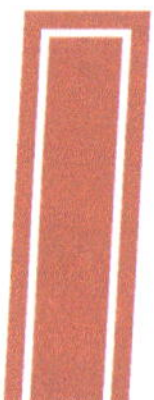

# Chapter-7
# Organofluorine chemistry

## 7. Organofluorine chemistry:

Organofluorine chemistry describes the chemistry of organofluorine compounds, organic compounds that contain the carbon–fluorine bond. Organofluorine compounds find diverse applications ranging from oil and water repellents to pharmaceuticals, refrigerants and reagents in catalysis. In addition to these applications, some organofluorine compounds are pollutants because of contributions to ozone depletion, global warming, bioaccumulation, and toxicity. The area of organofluorine chemistry often requires special techniques associated with the handling of fluorinating agents.

Fluorine, the most electronegative of all the elements, forms very strong covalent (see table 1) or ionic bonds to most other elements. The strength of the carbon-fluorine bond and the small size of the fluorine atom (van der Waals radius: 1.35 Å; hydrogen: 1.20 Å) give rise to a range of valuable chemical, physical and biological properties in organic molecules with one or more fluorine atoms attached to carbon. A rapid growth of interest in fluoro-organics has occurred in many areas of application, including polymers and materials, specialty solvents, performance fluids, medicinal agents, agrochemicals and in numerous reagents and intermediates for chemical synthesis.

**BOND BOND ENERGY**

| | kcal $mol^{-1}$ | kJ $mol^{-1}$ |
|---|---|---|
| F-F | 38 | 159 |
| Cl-Cl | 58 | 242 |
| H-F | 136 | 566 |
| H-Cl | 103 | 431 |
| C-H | 98 | 411 |
| C-F | 116 | 484 |
| C-Cl | 81 | 338 |
| Si-F | 139 | 582 |
| Si-Cl | 91 | 381 |
| P-F | 117 | 490 |
| P-Cl | 76 | 319 |

***Table 1: Typical covalent bond energies***

## 7.1 The carbon–fluorine bond:

Fluorine has several distinctive differences from all other substituents encountered in organic molecules. As a result, the physical and chemical properties of organofluorines can be distinctive in comparison to other organohalogens.

1. The carbon–fluorine bond is one of the strongest in organic chemistry (an average bond energy around 480 kJ/mol[1]). This is significantly stronger than the bonds of carbon with other halogens (an average bond energy of e.g. C-Cl bond is around 320 kJ/mol[1]) and is one of the reasons why fluoroorganic compounds have high thermal and chemical stability.
2. The carbon–fluorine bond is relatively short (around 1.4 Å[1]).
3. The Van der Waals radius of the fluorine substituent is only 1.47 Å,[1] which is shorter than in any other substituent and is close to that of hydrogen (1.2 Å). This, together with the short bond length, is the reason why there is no steric strain in polyfluorinated compounds. This is another reason for their high thermal stability. In addition, the fluorine substituents in polyfluorinated compounds efficiently shield the carbon skeleton from possible attacking reagents. This is another reason for the high chemical stability of polyfluorinated compounds.
4. Fluorine has the highest electronegativity of all elements: 3.98.[1] This causes the high dipole moment of C-F bond (1.41 D[1]).
5. Fluorine has the lowest polarizability of all atoms: 0.56 $10^{-24}$ $cm^3$.[1] This causes very weak dispersion forces between polyfluorinated molecules and is the reason for the often-observed boiling point reduction on fluorination as well as for the simultaneous hydrophobicity and lipophobicity of polyfluorinated compounds whereas other perhalogenated compounds are more lipophilic.

| Element | Van der Waals radius (Å) | Electro-negativity |
|---|---|---|
| F | 1.47 | 3.98 |
| O | 1.52 | 3.44 |
| N | 1.55 | 3.04 |
| C | 1.70 | 2.55 |
| H | 1.20 | 2.2 |

***Table 2: Vander Waals radius of electronegative element***

In comparison to aryl chlorides and bromides, aryl fluorides form Grignard reagents only reluctantly. On the other hand, aryl fluorides, e.g. fluoroanilines and fluorophenols, often undergo nucleophilic substitution efficiently.

## 7.2 Methods for preparation of C-F bonds:

Organofluorine compounds are prepared by numerous routes, depending on the degree and regiochemistry of fluorination sought and the nature of the precursors. The direct fluorination of hydrocarbons with $F_2$, often diluted with $N_2$, is useful for highly fluorinated compounds:

$$R_3CH + F_2 \rightarrow R_3CF + HF$$

Such reactions however are often unselective and require care because hydrocarbons can uncontrollably "burn" in $F_2$, analogous to the combustion of hydrocarbon in $O_2$. For this reason, alternative fluorination methodologies have been developed.

Because of the reactivity and hazards of elemental fluorine and hydrogen fluoride, the task of introducing fluorine into organic molecules has presented a particular challenge to synthetic chemists and has led to the development of specialized fluorination technologies and reagents. This article gives a brief outline of fluorination methods, highlighting specific reagents available from market.

### 7.2.1 Source of Fluorine:

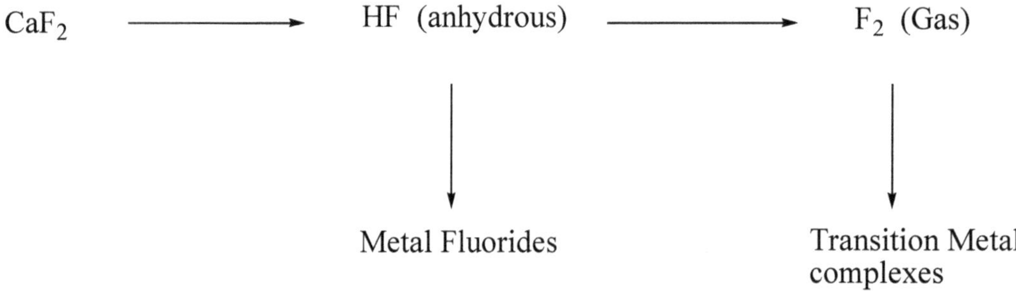


Fluorine, $F_2$, is a diatomic molecule existing as a pale yellow gas. It liquefies at -188 °C to produce a yellowish orange liquid, and solidifies at -220 °C to give a yellow solid. Its name derives from the Latin verb "fluere"(to flow) that explains the given name to fluorite (fluorspar) since $CaF_2$ exhibits good fluxing abilities.

Fluorine is the most reactive element, and the most powerful oxidizing element. It readily reacts with almost all organic and inorganic materials.

## 7.3 Fluorinating reagents:

Here we describe some fluorinating reagents for fluorination in organic compounds:

### 7.3.1. Sulfur fluorides

Sulfur tetrafluoride is a powerful fluorinating agent which has the disadvantage of being a highly toxic, corrosive gas.

Deoxofluorination agents effect the replacement hydroxyl and carbonyl groups with one and two fluorides, respectively. One such reagent, useful for fluoride for oxide exchange in carbonyl compounds, is sulfur tetrafluoride:

$$RCO_2H + SF_4 \rightarrow RCF_3 + SO_2 + HF$$

It converts alcohols to alkyl fluorides (Scheme 1) and carboxyl groups to trifluoromethyl (Scheme 1).[2,3]

**Scheme 1**

$$R{-}OH \xrightarrow{SF_4} R{-}F$$

$$RC(=O)OH \xrightarrow{SF_4} RCF_3$$

$$RC(=O)H \xrightarrow{SF_4} RCF_2H$$

Diethylaminosulfur trifluoride (DAST), a reagent derived from $SF_4$, is more convenient to handle and more selective, and is particularly useful for conversion of alcohols to alkyl fluorides, carboxylic acids to acyl fluorides (Scheme 2) and carbonyl compounds to gem-difluorides (Scheme 2). DAST has limited thermal stability at elevated temperatures, and other reagents have been developed, including Diethylaminosulfur trifluoride (DAST, $NEt_2SF_3$), Morpholinosulfur trifluoride ("Morpho- DAST") and bis(2-methoxyethyl)-aminosulfur trifluoride (deoxo-fluor) with improved thermal characteristics. These organic reagents are easier to handle and more selective:[4]

**Scheme 2**

$$R{-}OH \xrightarrow{Et_2NSF_3} R{-}F$$

$$RC(=O)OH \xrightarrow{Et_2NSF_3} RC(=O)F$$

$$RC(=O)R_1 \xrightarrow{Et_2NSF_3} RCF_2R_1$$

Fluorination with DAST and related reagents has been reviewed by Hudlicky.[5]

### 7.3.2 N-Fluoro reagents:

A variety of N-fluorinated amines, quaternary salts, amides and sulfonamides have been proposed as reagents for selective electrophilic fluorination under mild conditions (scheme 3). These are usually stable, easily-handled solids, and provide a range of fluorinating power from mild to moderate, depending on the structure of the reagent and the nature of the substrate. Another advantage is that, since free hydrogen fluoride is not a major by-product in fluorination reactions with this type of reagent, conventional glass equipment is often suitable.

**Scheme 3**

$$\equiv N^{\oplus}{-}F \; + \; {-}C{-}H \longrightarrow \; \equiv N^{\oplus}{-}H \; + \; {-}C{-}F$$

The chemistry of N-F fluorinating agents has been reviewed.[6]

- **a.** 1-Chloromethyl-4-fluoro-1, 4-diazoniabicyclo [2.2.2]octanebis(tetrafluoroborate) [F-TEDA-BF4]
- **b.** N-Fluorobenzenesulfonimide [Accufluor® NFSi]
- **c.** 1-Fluoro-4-hydroxy-1,4-diazoniabicyclo [2.2.2] octane bis(tetrafluoroborate) [Accufluor® NFTh]
- **d.** N-Fluoropyridiniumpyridineheptafluoro- diborate[Accufluor®NFPy]
- **e.** N-Fluoropyridinium trifluoromethane sulfonate

### 7.3.3 Nucleophilic reagents:

Fluoride ion is normally the least nucleophilic of the halides. Nevertheless, displacement of other halogens in alkyl halides can be effected, since the high stability of alkyl fluorides and the poor leaving group ability of $F^-$ can cause the equilibria to be shifted (Scheme 4).

**Scheme 4**

$$-\overset{|}{\underset{|}{C}}-X \; + \; F^{\ominus} \; \rightleftharpoons \; -\overset{|}{\underset{|}{C}}-F \; + \; X^{\ominus}$$

Dipolar aprotic solvents, such as DMF or acetonitrile, tend to give the best results, and in view of the low solubility of metal fluorides, addition of a crown ether can be beneficial; alternatively, the much more soluble tetraalkylammonium fluorides can be employed. In aromatic systems, displacement of chloride (halex fluorination) can be achieved in high-boiling polar aprotic solvents including DMSO or sulfolane (Scheme 5).

**Scheme 5**

R–C₆H₄–Cl —(KF, Sulfolane / Heating)→ R–C₆H₄–F

The commonest fluoride source is potassium fluoride, though other fluorides are sometimes used, and improved results can often be obtained if the fluoride ion is solubilised by means of a thermally stable phase-transfer catalyst such as tetraphenyl phosphonium chloride. The aromatic nucleophilic substitution reactions of fluoride have been reviewed by Vlasov.[7] Inorganic fluorides also have a variety of other uses in synthesis,[8] for example as mild bases in condensation reactions. Inorganic and tetraalkylammonium fluorides are widely used in the selective cleavage of silylderivative.[9]

- **a.** Calcium fluoride
- **b.** Cesium fluoride
- **c.** Lithium fluoride
- **d.** Potassium fluoride
- **e.** Silver(I) fluoride
- **f.** Sodium fluoride

### 7.3.4 Hydrogen fluoride and equivalents:

Although hydrogen fluoride may appear to be an unlikely nucleophile, it is the most common source of fluoride in the synthesis of organofluorine compounds. Such reactions are often catalysed by metal fluorides such as chromium trifluoride. 1,1,1,2-Tetrafluoroethane, a replacement for CFC's, is prepared industrially using this approach:[10]

$$Cl_2C{=}CClH + 4\,HF \rightarrow F_3CCFH_2 + 3\,HCl$$

Notice that this transformation entails two reaction types, metathesis (replacement of $Cl^-$ by $F^-$) and hydrofluorination of an alkene.

**Electrochemical fluorination** This technique was introduced by J. H. Simons in the 1940's, using anhydrous HF as solvent, with or without an ionic fluoride as supporting electrolyte, and is now an established industrial route to perfluorinated molecules. More recent studies have investigated the selective fluorination of a variety of aliphatic and aromatic substrates, for which acetonitrile has become the preferred solvent, with the addition of a supporting electrolyte such as HF-pyridine or triethylamine trihydrofluoride.[11]

- **a.** Hydrogen fluoride pyridine complex
- **b.** Potassium hydrogen fluoride
- **c.** Triethylamine trihydrofluoride
- **d.** Tetraethylammonium tetrafluoroborate

Hydrogen fluoride is a relatively weak acid (pKa= 3.45 at 25 °C), providing only low concentrations of fluoride ions in solution. Combined with the comparatively low nucleophilicity of fluoride, this tends to make HF a rather ineffective fluorinating agent, for example in the conversion of C-OH bonds to C-F. Furthermore, anhydrous HF is very volatile (b.p. 20 °C), highly toxic, extremely corrosive to skin and other tissues including bone, and readily attacks glass. Many other reagents, formally equivalent to HF are available for the introduction of fluorine in specific molecular environments. **HF-amine complexes** These are either liquids which are less volatile and easier to handle than HF itself, or crystalline solids. They tend to be more nucleophilic than HF, making them valuable reagents for various types of fluorination. (See also use in electrochemical fluorination, above). The most frequently used of these reagents is HF-pyridine (Olah's Reagent), applications of which include: preparation of alkyl fluorides from alcohols (schemes 5) or alkenes (scheme 6); acyl fluorides from acyl chlorides or anhydrides; anddeaminative fluorination of amino acids and arylamines (Scheme 6).

**Scheme 6**

$$R{-}OH \xrightarrow{(Py).(HF)x} R{-}F$$

$$C{=}C \xrightarrow{(Py).(HF)x} H{-}C{-}C{-}F$$

$$R{-}CH(NH_2){-}COOH \xrightarrow[(Py).(HF)x]{NaNO_2} R{-}CH(N^{\oplus}{\equiv}N)(COOH)\ F^{\ominus} \longrightarrow R{-}CH(F){-}COOH + N_2$$

The use of HF-amine complexes as fluorinating agents has been reviewed byYoneda.[12]

- **a.** Hydrogen fluoride 2,4,6-collidinecomplex
- **b.** Hydrogen fluoride pyridine complex
- **c.** Triethylamine trihydrofluoride
- **d.** Tetra-n-butylammonium dihydrogen- trifluoride, 50-55% w/w in 1,2-dichloroethane

### 7.3.5 Electrophilic reagents:

Xenon fluorides, especially the difluoride, can be used in the selective fluorination of substrates including arenes, alkenes and activemethylenes, and also in the fluoro decarboxylation of carboxylic acids (scheme 7), providing a useful alternative to the Hunsdiecker reaction.[13]

**Scheme 7**

$$R{-}C(=O){-}OH \xrightarrow{XeF_2} R{-}F + CO_2 + Xe$$

The use of cobalt(III) fluoride [or cobalt(II) fluoride/ fluorine] is a well-established method for the fluorination of hydrocarbons, ethers, etc., but requires high temperatures and specialised equipment. Silver(II) fluoride has found some limited use in the fluorination of aromatics. Several highly reactive species can be generated from elemental fluorine, including trifluoromethyl hypofluorite, acetyl hypofluorite (explosive), and cesium fluoroxy- sulfate (unstable, shock-sensitive), prepared from cesium sulfate and fluorine.

- **a.** Cobalt(II) fluoride
- **b.** Cobalt(III) fluoride
- **c.** Silver(II) fluoride
- **d.** Xenon difluoride

**Antimony fluorides** The use of antimony fluorides to displace other halogens has played a major role in the development of the fluorocarbon industry. The reagents can be used either in stoichiometric amounts or catalytically in the presence of HF. The latter method was widely employed in the synthesis of chlorofluorocarbons (CFCs) and hydrochlorofluorocarbons (HCFCs). Further information can be found in relevant sections of general works on fluorination.[14,15]

- **a.** Antimony(III) fluoride
- **b.** Antimony(V) fluoride

### 7.3.6 Fluoroborates and analogues:

The classical Balz-Schiemann (or Schiemann) conversion of arylamines to aryl fluorides (Scheme 8) involves the formation and isolation of the diazonium tetrafluoroborate, followed by thermolysis either neat or in an inert solvent.[16,17]

**Scheme 8**

$$R{-}NH_2 \xrightarrow[HBF_4]{NaNO_2} R{-}\overset{\oplus}{N}{\equiv}N\ \overset{\ominus}{B}F_4 \xrightarrow{Heating} R{-}F$$

Direct formation of aryldiazonium tetrafluoroborates from arylamines and nitrosonium tetrafluoroborate in organic solvents gives good yields of aryl fluorides. Improved results may also be obtained by the use of the diazonium hexafluorophosphate, or a one-pot direct diazotization in anhydrous HF or HF-pyridine.

- **a.** Tetrafluoroboricacid
- **b.** Hexafluorophosphoricacid
- **c.** Hydrogen fluoride pyridine complex
- **d.** Nitrosonium tetrafluoroborate
- **e.** Silver tetrafluoroborate

### 7.3.7 Other reagents:

For the conversion of alcohols to alkyl fluorides and carboxylic acids to acylfluorides, Ishikawa's reagent (N,N-diethyl-1,1,2,3,3,3-hexafluoropropylamine/ N,N-diethyl-1,2,3,3,3- pentafluoropropenamine) has been found effective. Cyanuricfluoride is a particularly valuable reagent, introduced by Olah, which converts carboxylic acids to acylfluorides under mild conditions.

**Scheme 9**

CbzHN–CH($R_1$)–COOH + 2,4,6-trifluoro-1,3,5-triazine $\xrightarrow[CH_2Cl_2]{Pyridine}$ CbzHN–CH($R_1$)–C(O)F $\xrightarrow{CbzHN–CH(R_2)–COOR_2}$ CbzHN–CH($R_1$)–C(O)–NH–CH($R_2$)–$COOR_2$

Its application to N-protected amino acids and use of the resulting acylfluorides in peptide synthesis (Scheme 9) has been developed byCarpino.[18]

a. Cyanuricfluoride
b. N,N-Diethyl(2-chloro-1,1,2-trifluoroethyl)- amine (Yarovenko'sReagent)
c. Ishikawa's Reagent
d. Tetrabutylammonium difluorotriphenyl- stannate(Gingras'Reagent)

### 7.4 Electrosynthetic methods:

A specialized but important method of electrophilic fluorination involves electrosynthesis. The method is mainly used to perfluorinate, i.e. replace all C–H bonds by C–F bonds. The hydrocarbon is dissolved or suspended in liquid HF, and the mixture is electrolyzed at 5–6 V using Ni anodes.[19] The method was first demonstrated with the preparation of perfluoropyridine ($C_5F_5N$) from pyridine ($C_5H_5N$). Several variations of this technique have been described, including the use of molten potassium bifluoride or organic solvents.

#### 7.4.1 Fluorine Generation:

Fluorine gas is generated by the electrolysis of anhydrous potassium bifluoride ($KHF_2$, or KF.HF) in HF.

The fluoride anion is oxidized at the anode to liberate $F_2$ gas.

$2F^- \rightarrow F_2 + 2e^-$ OILRIG

At the cathode, protons are reduced to hydrogen gas.

$2H^+ + 2e^- \rightarrow H_2$

(Best to make sure the two compartments are kept separate!)

Anhydrous HF has a very low electrical conductivity, and so cannot be used as the electrolyte by itself. That's why the molten fluoride salt electrolyte is used.

There are a few compounds that can release $F_2$ when heated or reacted –but those compounds were originally made using $F_2$.

A chemical route, which does not rely on compounds derived from $F_2$ was reported by Karl Christein 1986.

$2\ K_2MnF_6 + 4\ SbF_5 \rightarrow 4\ KSbF_6 + 2\ MnF_3 + F_2$

The starting materials are prepared from HF, and at 150 °C they react to liberate Fluorine gas.

**7.5 fluorinated building blocks:**

Many organofluorine compounds are generated from reagents that deliver perfluoroalkyl and perfluoroaryl groups. (Trifluoromethyl)trimethylsilane, $CF_3Si(CH_3)_3$, is used as a source of the trifluoromethyl group, for example.[20] Among the available fluorinated building blocks are $CF_3X$ (X = Br, I), $C_6F_5Br$, and $C_3F_7I$. These species form Grignard reagents that then can be treated with a variety of electrophiles. The development of fluorous technologies (see below, under solvents) is leading to the development of reagents for the introduction of "fluorous tails."

A special but significant application of the fluorinated building block approach is the synthesis of tetrafluoroethylene, which is produced on a large-scale industrially via the intermediacy of difluorocarbene. The process begins with the thermal (600-800 °C) dehydrochlorination of chlorodifluoromethane:[21]

$$CHClF_2 \rightarrow CF_2 + HCl$$

$$2\ CF_2 \rightarrow C_2F_4$$

Sodium fluorodichloroacetate (CAS# 2837-90-3) is used to generate chlorofluorocarbene, for cyclopropanations.

**7.6 $^{18}$F-Delivery methods:**

The usefulness of fluorine-containing radiopharmaceuticals in $^{18}$F-positron emission tomography has motivated the development of new methods for forming C-F bonds. Because of the short half-life of $^{18}$F, these syntheses must be highly efficient, rapid, and easy.[22] Illustrative of the methods is the preparation of fluoride-modified glucose by displacement of a triflate by a labeled fluoride nucleophile:

## 7.7 References:


1. Kirsch, Peer *Modern fluoroorganic chemistry: synthesis, reactivity, applications*. Wiley-VCH, 2004.
2. 'Fluorination bysulfur tetrafluoride', G. A. Boswell, W. C. Ripka, R. M. Scribner, C. W. Tullock, *Org. React.,* **21**, 1 (1974).
3. 'Fluorination by sulfur tetrafluoride', C.-l. J. Wang, *Org. React.,* **34**, 319 (1985).
4. Gauri S. Lal, Guido P. Pez, Reno J. Pesaresi and Frank M. Prozonic (1999). "Bis(2-methoxyethyl)aminosulfur trifluoride: a new broad-spectrum deoxofluorinating agent with enhanced thermal stability". Chemical Communications (2): 215.
5. 'Fluorination with diethylaminosulfur trifluoride and related aminofluoro-sulfuranes', M. Hudlicky, *Org. React.,* **35**, 513 (1988).
6. 'Electrophilic N-fluorinating agents', G. S. Lal, G. P. Pez, R. G. Syvret, *Chem. Rev.,* **96**, 1737 (1996).
7. 'Fuorideion as anucleophile and a leaving group in aromatic nucleophilic substitution reactions', V. M. Vlasov, *J Fluorine Chem.,* **61**, 193 (1993).
8. 'Alkali metal fluorides in organic synthesis', G. G. Yakobson, N. E. Akhmetova, *Synthesis*, 169 (1983).
9. 'Fluoride ion as a base in organic synthesis', J. H. Clark, *Chem. Rev.*, **80**, 429 (1980).
10. William R. Dolbier, Jr. (2005). "Fluorine Chemistry at the Millennium". Journal of Fluorine Chemistry 126 (2): 157. doi:10.1016/j.jfluchem. 2004.09.033.
11. 'Review of rccent developments in the selective electrochemical fluorination of organic compounds', M. Noel, V. Suryanarayanan, S.Chellamal, J. FluorineChem., 83, 31 (1997).
12. 'The combination of hydrogen fluoride with organic bases as fluorinating agents', N.Yoneda, Tetrahedron, 47, 5329 (1991).
13. 'Xenondifluoride in synthesis', M. A. Tius, *Tetrahedron,* **51**, 6605 (1995).
14. *Chemistry of Organic Fluorine Compounds*, M. Hudlicky, Ellis Horwood, Chichester (1976).
15. 'The preparation of aliphatic fluorine compounds', A. L. Hesse, *Org. React.,* **2**, 49 (1944).
16. 'Preparation of aromatic fluorine compounds from diazonium fluoborates: the Schiemann reaction', A. Roe, *Org. React.,* **5**, 193 (1949).
17. 'The Balz-Schiemann reaction', H.Suschitzky, *Adv. Fluorine Chem.,* **4**, 1 (1965).
18. 'Peptide synthesis via amino acid halides', L. Carpino, *Acc. Chem. Res.,* **29**, 268 (1996).

19. J. H. Simons "The Electrochemical Process for the Production of Fluorocarbons" Journal of The Electrochemical Society, 1949, Volume 95, pp. 47-66. doi: 10.1149/1.2776733
20. Pichika Ramaiah, Ramesh Krishnamurti, and G. K. Surya Prakash (1998), "1-trifluoromethyl)-1-cyclohexanol", *Org. Synth.*: 232
21. *Chemistry of Organic Fluorine Compounds*, M. Hudlicky, Ellis Horwood, Chichester (1976).
22. Le Bars, D. (2006). "Fluorine-18 and Medical Imaging: Radiopharmaceuticals for Positron Emission Tomography". *Journal of Fluorine Chemistry* **127** (11): 1488–1493. doi:10.1016/j.jfluchem. 2006.09.015.

www.ingramcontent.com/pod-product-compliance
Lightning Source LLC
Chambersburg PA
CBHW050727180526
45159CB00003B/1148
* 9 7 8 1 5 1 1 7 4 1 6 6 8 *